THE BILINGUAL BRAIN

The Bilingual Brain

Arturo E. Hernandez

OXFORD
UNIVERSITY PRESS

OXFORD
UNIVERSITY PRESS

Oxford University Press is a department of the University of Oxford.
It furthers the University's objective of excellence in research, scholarship,
and education by publishing worldwide.

Oxford New York
Auckland Cape Town Dar es Salaam Hong Kong Karachi
Kuala Lumpur Madrid Melbourne Mexico City Nairobi
New Delhi Shanghai Taipei Toronto

With offices in
Argentina Austria Brazil Chile Czech Republic France Greece
Guatemala Hungary Italy Japan Poland Portugal Singapore
South Korea Switzerland Thailand Turkey Ukraine Vietnam

Published in the United States of America by
Oxford University Press
198 Madison Avenue, New York, NY 10016

Library of Congress Cataloging-in-Publication Data
Hernandez, Arturo E.
The Bilingual Brain / Arturo E. Hernandez.
pages cm
Includes bibliographical references and index.
ISBN 978–0–19–982811–1
1. Bilingualism–Psychological aspects. 2. Bilingualism—Physiological aspects.
3. Code switching (Linguistics) 4. Psycholinguistics. I. Title.
P115.4.H48 2013
404′.2019—dc23 2012050204

9 8 7 6 5 4 3 2 1
Printed in the United States of America
on acid-free paper

To my mother, Lucha Corpi, for teaching me two languages,
my father, Guillermo Hernandez, for planting the seeds for two more,
and Elizabeth Bates, for helping me to make sense of it all.

Contents

Acknowledgments

WRITING A BOOK is a long and arduous journey. I really had no appreciation for how difficult it was until I undertook it myself. First and foremost, I have to thank my wife, Naomi, for supporting me during this tumultuous process. Her unwavering faith, grounded outlook on life, and realism always served to keep me on the right path. I always joke that if I got stranded on an island and I had to choose one person to accompany me it would be her and not because she is my wife. She is able to keep a cool and level head in any situation. Her words of advice and support were always there

I also have to thank my mother, Lucha. An accomplished writer is the best role model, and my mother provided that in ample supply. From rising early in the morning to the importance of telling a story, my mom had advice for every possible contingency. Of course, this has nothing to do with her bringing me to this world, caring for me when I was sick, or buying me my first baseball glove. My mom did all that but most importantly she fed me, clothed me, and paraded me through a literary world ever since I can remember.

I also have to give thanks to two people who left this world much too early. The first is my father, Guillermo, whose optimism, humor, and sense of wonder opened a lot of doors in my mind. He was the one who always saw the future and wondered what things might be like. His vision and ability to look past the words to see the subtext were invaluable. For him, there were always multiple levels to any piece of literature. We often talked about a whole host of things, including politics, economics, technology, and his favorite, Mexican folk music. But most important was the incredible belief he had in me. He often said I was his hero. I could always find solace in his voice and his encouraging words. He was always there (if not in person at least in spirit) even when things seemed hopeless and the finish line seemed so out of reach.

And if it was not enough to have all these valuable inspirations next to me, I was also fortunate to be trained by one of the most brilliant cognitive neuroscientists and psycholinguists that ever graced the planet, Elizabeth Bates. Liz, as we all knew her, was one of the most articulate and inspirational people I ever met. In the old times of Unix terminals and telnet windows, Liz frequently froze her computer because she typed so quickly. She once sat down to write a review of a paper and stopped 2 hours later with 25 pages in hand. But her extreme talent was never used to make her students less. She taught me that graduate students were the most important beings in a university. Her students were like adopted children. She would take us out to lavish meals (at least for graduate students), introduce us to all the grown ups that came parading through the lab, and even give us (me actually) free tickets from unused miles she accumulated from her frequent trips. Thanks to her I got to travel to cities I had never visited and to countries that I never thought I would see. But aside from her support and all the opportunities that she gave me, I had a true intellectual affinity with Liz. She and I both thought of nature as fluid and movement oriented. There were no fixed properties but rather probabilities and states of being. The six years I spent at the University of California, San Diego, were some of the most transformational years in my life, and I have Liz to thank for socializing me and letting me know that anything was possible. At times I still catch a glimpse of her influence on me. It might be in the way I phrase something in talks that I give or in throwaway lines that manage to permeate the pages when I am writing. Like everyone she ever worked with, I miss her enormously.

Many other people also made this book possible. This includes my three children, Kiara, Nikolas, and Kamille, who have all grappled with bilingualism in its various forms. Observing them often led to little insights that helped to further my thinking about this topic. In addition, they found ways to support my endeavor. They babysat, made lunch for themselves, and left me alone so I would have time to write. I would also like to thank all of my present and former graduate students, Gayane Meschyan, Lynette Austin, Kelly King, Lindsay Carr, Eric Waldron, Pilar Archila-Suerte, Elizabeth Woods, Kailyn Bradley, Aurora Ramos, Vicki Wagner, and Maya Ravid. Without their efforts in tackling the day-to-day work needed to keep a lab running, it would have been impossible for me to get everything done and still find time to write this book. In addition, I would like to thank the National Institute of Child Health and Human Development (NIH R21HD059103–01 and 3R21HD059103–02S1) and the von Humboldt Foundation for their financial support of the research that is included in the book.

I would like to thank the large number of collaborators that have contributed in various ways to this book. This includes members of my doctoral committee, Jeff Elman, Javier Movellan, and Marta Kutas. The three of them served to inspire me and prod me into doing more. Others who helped me include Susan Bookheimer, Mirella Dapretto, and John Mazziotta at the University of California, Los Angeles. Angela Friederici and Sonja Kotz welcomed me twice to the Max Planck Institute in Leipzig. Christian Fiebach and Chrisoph Herrmann helped ease the transition into German life and provided me with insight into research on and beyond language. Isabell Wartenburger and the research group at the Charite in Berlin allowed me to take the first glimpses of the child bilingual brain in action. Jason Zevin and Joe Devlin have taken on the role of neural network modelers turned cognitive neuroscientists. I would like to thank Ping Li, whom I consider my intellectual cousin. Liz Bates always thought that I should talk more to him, and over the years I keep remembering that she was right. He and I have collaborated on a pair of more theoretical projects that have helped to frame many of the issues that are discussed in this book.

I would like to thank my colleagues David Francis and Jack Fletcher, who were able to read or reflect on various aspects and encouraged me to write this in lieu of several papers or a grant or two. Other members of the developmental cognitive neuroscience area, Bruno Breitmeyer, Donald Foss, Leigh Leasure, Merrill Hiscock, and Hanako Yoshida, engaged me in many spirited conversations that helped to further refine many of the issues in this book.

I would like to thank several people in the literary world. This includes Susan Rabiner, who spoke to me by phone and guided me toward a more academic press. Phil Laughlin at MIT Press, who spent time talking to me about the book and actually got me to realize where the book began and ultimately how to frame it. And, of course, I have to thank Joan Bossert at Oxford University Press for providing the guidance and lending a helping hand in giving the text its final form. This included helping me sift through two anonymous reviews that provided substantial feedback on an earlier version of this manuscript.

Writing can be an extremely lonely process. All of those who helped in some way served to make it just a little bit less lonely.

Arturo E. Hernandez
July 24, 2012
Houston, TX

1

ANTAGONISTS, PARASITES, AND SYMBIOTES

IT WAS AT the age of 20 that I first encountered the paradox that would come to dominate my academic life and that serves as the major impetus for this book. I was taking a class in cognitive neuropsychology at the University of California, Berkeley, with Art Shimamura, a newly minted assistant professor, who had moved up the coast from La Jolla after a postdoc with Larry Squire. During class we learned about patients who had suffered strokes and lost the ability to do things that most people take for granted. Some lost the ability to remember new things (amnesia), some lost the ability to perceive things on the left side of space (spatial neglect), and others lost the ability to use language (aphasia).

But the class unit on language actually began with a discussion of presurgical mapping. The pioneering work by Wilder Penfield in the 1930s (Penfield & Boldrey, 1937) had been extended by George Ojemann, a neurosurgeon at the University of Washington Medical School. In this work neurosurgeons stimulate the brain directly in order to map out the language zones. Based on this literature, Kolb and Whishaw (1985) suggested that individual variability in the size of the language zones was most likely due to how well a person used the language. Two facts supported this view. First, females had smaller language zones than males, a likely reflection of their superior verbal skills according to the authors at the time. They suggested that less ability in a language results in a more diffuse brain representation, which "is supported by his [Ojemann's] observation that in multilinguals the poorer language is disrupted over a larger area than the better one" (Kolb & Whishaw, 1985, pp. 570–571). In short, there seemed to be a relationship between how focally a language is represented in the brain and how well it is spoken. Surprisingly, the entire chapter we read on language contained no additional reference to the neural representation of bilingualism or multilingualism.

My curiosity piqued by this topic, I made an appointment to see my professor. Art Shimamura was personable and approachable and he patiently sat with me as I asked him about bilinguals. He pointed me to the work of Ojemann that had been mentioned in the class textbook (Ojemann & Whitaker, 1978). I quickly went to the library and looked the article up only to find that the two cases presented showed a very different profile. There was no place in the brain for each language. Furthermore, the patients had very different language profiles. One was a late bilingual, whereas the other was an early bilingual.

Intrigued by this topic, I spent time in the library hunting through the aphasia literature looking for cases of bilinguals who had lost the ability to use one of their languages after some form of brain damage. I asked a teaching assistant, Felicia Gershberg, whether she thought it would be appropriate to do my final paper on bilingual aphasia. She said yes. Luck would have it that there was a special issue of *Brain and Language* dedicated to this topic.

As I read through the different articles, I stumbled across a case that caught my attention. A.S. was a 49-year-old experienced orthopedic surgeon who spoke Farsi as a native language but went to college in Germany at age 18. During this time, he also spent a year doing research in England. Although Farsi was his native language, the patient also spoke German and English, his second and third languages.

On September 2, 1985, at the age of 49, A.S. was injured by an explosion that led to left frontotemporal trauma. He lost consciousness temporarily and was hospitalized that same day. After regaining consciousness he could say a few words in Farsi for the first 16 days before he switched to German. Much to everyone's surprise, he could understand both Farsi and German, but could only speak in German even to monolingual Farsi speakers for the next 3 weeks, after which he switched back to Farsi. As he alternated between German and Farsi during these first few weeks, he was unable to speak or understand English. He regained the use of English only after having completely recovered his mastery of both Farsi and German.

Why did this particular case study capture my interest? Just two years earlier, I had experienced a similar episode during my first few months in Brazil, where I was immersed for the first time in Portuguese, my third language. In a brief period of time, speaking Spanish, my mother tongue, became extremely difficult and labored. In addition, English, my dominant language, was quickly becoming less and less accessible. I found it very difficult to write, read, or speak it with the fluency I had possessed just six months earlier. Over two years, I became so proficient in Portuguese that people mistook me for a native. But my newfound proficiency in Portuguese came with a cost. My return from Brazil was marked with a couple of semesters of agony as I slogged through textbooks that used to be a breeze. And my Spanish, which had been native-like, now sounded strange to my grandmother. At the age of 20, like patient A.S., I had suffered my own alternating antagonism, during which increased accessibility in one language was accompanied by loss in another language. But this interference between languages was not due to brain damage. It was simply the product of being immersed in a third language.

Cases of language loss and recovery bring up an intriguing paradox. If two languages are stored in the brain, how can it be that a person could lose a language and gain it back? My own experience and the cases of bilingualism that I was reading at the time suggested a very different model of how a language was represented in the brain. It appeared that languages could be inaccessible even though they were not entirely lost. Hence, stress, whether it be due to foreign language immersion, sleep deprivation, or brain damage, could lead to the apparent loss of one language and not the other. In my case, even though Spanish was my mother tongue, it was a language that I had used to a much lesser extent. English was my most literate language and as such it withstood the onslaught of Portuguese immersion much better.

This episode also showed me that the brain was not egalitarian. Rather than giving similar representational privileges to all languages, it favored some languages over others. And so I set off on a quest to understand the factors that might help to predict how two languages are coded in one brain. This book fleshes out this topic with results that have come out in the 22 years since I first stumbled on this field of study as an undergraduate at the University of California, Berkeley. In this time researchers spanning the globe have generated a lot of new information. However, interestingly a lot this information builds on existing notions that were introduced many years ago.

FAMILIARITY AND ITS NEURAL INSTANTIATION

The phenomenon of alternating between languages shown by patient A.S. (and by me) shares the flavor of the conclusion that Kolb and Whishaw suggested in the third edition of their classic textbook. Stronger languages are more efficient and hence may be most resistant to brain damage. The weaker language may be more spread out across the cortex and hence may be less resistant to damage.

The idea that stronger languages might be more resistant to damage was also present in early work carried about by Pitres in 1895. Jean-Albert Pitres was born in Bordeaux in 1848 (Lebrun, 1995). He studied medicine in his birthplace before joining the army and serving as an auxiliary physician in Paris. He completed his studies in Paris and went on to work with Charles Emile Francois-Frank (1849–1921). Upon completing his training he returned to Bordeaux, where he took a position in the newly formed school of medicine.

Pitres noted that selective impairment of a language in a patient who was bilingual or multilingual was quite common. He then went on to review a number of cases in which he describes the loss of language in multilingual aphasics. Some patients retain access to only one language. For example, consider the case of a 56-year-old male who lived in France but learned German as a child and Basque, English, Spanish, and Italian as an adult. After a suspected stroke during which he lost consciousness, he had difficulty moving the right side of his body, what neurologists would call right-sided hemiplegia. Over time he recovered his ability to use French but not his other five languages. Although other cases reviewed by Pitres showed recovery in several languages, there was always a differential amount of recovery of some languages relative to others.

Based on these cases, Pitres formulated the generalization that the strongest language would remain resistant to damage. This is known as Pitres's rule. In addition, Pitres noted that many cases showed temporary "inertia" and that once this was overcome there was no need to think that aphasics were growing new language centers for each language.

The idea that familiarity of language is important for the recovery of language has played an important role in the child and adult bilingual literature. Language proficiency, as it has been termed today, has been found to play a central role in the neural representation of bilingualism (Abutalebi, Cappa, & Perani, 2001; Chauncey, Midgley, Grainger, van Heuven, & Holcomb, 2004; Chee, Hon, Lee, & Soon, 2001; Elston-Guettler, Paulmann, & Kotz, 2005; Mechelli et al., 2004; Meschyan & Hernandez, 2006; Perani et al., 1998; van Heuven, Dijkstra, & Grainger, 1998; Xue, Dong, Jin, Zhang, & Wang, 2004) and in the behavioral responses seen in the cognitive literature (Birdsong, 2006; Costa, Santesteban, & Ivanova, 2006; Kotz & Elston-Guettler, 2004). To this day, the idea that stronger languages are represented differently than weaker languages remains.

AGE OF ACQUISITION

In addition to language proficiency, the age at which a language was learned, age of acquisition (AoA), can also be seen as playing a role in the recovery of function observed with patient A.S. The alternating antagonism he experienced was between German and Farsi, his first and second language. The third language,

English took the longest to recover. This case suggests that when a language is learned may also play a role in its speed of recovery. This can be seen in some case studies. The question remains whether other cases show the importance of AoA in their recovery.

Other cases show a similar resistance to damage of earlier languages (Bianchi, 1886). There is the case of A.A., a healthy 24-year-old man, short and muscular but well proportioned. He was an excellent businessperson working successfully as a traveling salesman abroad. Aside from Italian, his native language, he also spoke French and English, which he had learned early in life. One day following an episode of extreme convulsions he was promptly taken to the doctor for treatment. Like A.S., A.A. suffered from hemiplegia, difficulty moving half his body, and had difficulty producing any speech. Upon examination it became clear that he could say some words in Italian but did not understand the link between these words and the corresponding objects in the real world. However, over time, the link between a word and its real-world referent became stronger. Unlike A.S., A.A. was never able to fully recover all the languages he spoke. Whereas he showed a considerable amount of recovery in Italian, he was never able to say a single word in French or English again. In the case of A.A., the first language was the last one left standing.

On the surface, the case of A.A. supports Pitres's rule, since Italian, the most familiar language, is preserved relative to the other two. But in this case, familiarity is confounded with AoA, since Italian was also the patient's first language. A.A. learned both French and English early and spoke both of them very well before the damage. However, they were learned after the first language was already in place. This would suggest that the first language might have some strength over and above pure familiarity. At least in this case, we would have to accept that AoA may not just be familiarity in disguise.

The notion that age of learning was a principal driving force in recovery of language for bilingual aphasics predates Pitres. Ribot, considered by many to be the father of French psychology, had proposed that language representation in bilinguals would be driven by when a language was learned (Lorch, 2009; Ribot, 1881, 1882). He saw this as a natural extension of the general laws that ruled the organization of memory. He noted that when patients had difficulties with word finding they would show the same order of loss: proper names, common nouns, adjectives and verbs, interjections, and finally gestures. In each of these categories, patients would forget their most recent memories first and their earliest memories last. This regression would proceed from items that were new, complex, voluntary, and least organized to items that were old, simple, automatic, and best organized. In short, Ribot proposed the idea that "age of acquisition" played a profound role in the formation of memories.

Ribot also proposed an interesting view on what today would be termed automaticity. He noted that there are very simple organic or motor memories, which are deeply engrained and not available to consciousness. This is what many today would term procedural memory (Squire & Zola, 1996). In addition, there are conscious memories that are formed over time. As we encounter a number of events across time, the time stamps associated with individual instances are lost. This would naturally lead to the loss of specifics for a particular episode and the retention of commonalities across memories. In short, these new conscious memories would take on the form of organic memories in that they would be deeply engrained. Furthermore, these new memories would involve the characteristics that were common across all the different episodes.

At the time that Ribot was developing his theories, there was little in the way of theoretical or empirical psychological work with children. However, there was an existing literature on second language acquisition that built on then-current ideas that distinguished between speech (motoric memory) and comprehension, which was considered a component of the mind. In fact, Ribot was hinting at the distinction between speech and other forms of memory. As we will see in this book, AoA seems to have effects on certain types of language functions, especially those related to speech. However, other components of the mind seem to be less influenced by AoA to a certain extent.

Finally, Ribot considers the case of the recovery of lost languages in multilinguals. Of particular interest to him were patients who under anesthesia showed amazing recovery of a language even after protracted periods of little use. One example that he uses is the case from Duval's article on hypnotism.

An old forester had grown up on the border with Poland, where he only spoke Polish. Afterward he lived in the German districts, and his children verified that for more than 30 years he neither heard nor spoke a single Polish word. However, upon being administered anesthesia he spent close to 2 hours speaking, praying, and singing, exclusively in Polish (Ribot 1881, 1882, p. 181).

The forester had these memories even though he had used German exclusively during his adult life. The deeply engrained memories were released as soon as the conscious mind was put to sleep. Hence, even though new memories can become very well engrained during an adult's life, the primary memories will remain over time when the conscious mind is put to sleep. The fact that the second language is built using the conscious mind and the first language is built using organic and motor memories is the basis for Ribot's claim that AoA plays a crucial role in the neural representation of both languages in a bilingual.

BEYOND LANGUAGE: THE ROLE OF CONTROL

Whereas Pitres and Ribot proposed the two mechanisms of AoA and familiarity/proficiency as playing a crucial role in the neural instantiation of two languages, there still exists the possibility that impairment may not have to do with damage to the cortex that is involved directly in language. Rather, the impairment may involve mechanisms that are involved in helping to select or switch to the correct language. Loss in the ability to switch from or to a language would result in the apparent loss of that language. However, it would not be loss per se but rather loss in the ability to control which language is available.

The first to propose the idea of a switching mechanism was Otto Pötzl, an Austrian neurologist born in 1877 in Vienna, Austria, where he completed his medical training. He is best known for his work during this tenure at the Vienna Psychiatric and Neurological Clinic, where he joined Freud's psychoanalytic movement and became one of its most ardent supporters. But our story takes us back to 1922, when he took a position as head of psychiatry in Prague (Pötzl, 1983). During this time he encountered a bilingual patient who had a strange syndrome.

The patient was a 60-year-old Austrian businessman, a native speaker of Czech, who had learned German at 14. As an adult he had moved to Vienna and used German exclusively in his family and daily life. The patient suffered two strokes that had caused him to lose the ability to see the space that was directly to his right. Furthermore, he suffered from a peculiar language impairment. He could understand both Czech and German but he could only speak Czech. A subsequent patient with a very similar problem led Pötzl to conclude that some bilinguals suffered from a problem with language fixation. Although they understood both languages perfectly, they remained fixated on the language they had used during their first few days after the stroke.

In 2000 Franco Fabbro (2000), one of the premier experts in the neurolinguistics of bilingualism, encountered a similarly odd case. A 56-year old Friulian man, patient S.I., was a prominent land surveyor who learned Friulian as a native language but was educated in Italian for 13 years. Although close to a million people speak this dialect in the northeast portion of Italy, the vast majority also speak Italian, the majority language. During a vacation with an Italian monolingual friend, the patient noticed that he began to intersperse Friulian sentences with his Italian friend. He kept apologizing and translating for his friend. But it was quite unnerving. Later he began to notice problems moving the right side of his body and eventually found walking more and more difficult.

The patient met with Fabbro and it was quickly determined that he had a tumor, which was resected. After the tumor had been removed, the patient presented with a very odd pattern of communication whereby he would alternate between his two languages across sentences. He never mixed languages within a sentence. But the patient would consistently say one sentence in Friulian and one in Italian even when he knew the person listening spoke only Italian. When tested with a set of standardized neuropsychological instruments, he showed no real language deficits. He understood each language very well. He could repeat words in each language almost perfectly. He could even translate between his languages. The problem seemed to be restricted to spontaneous speech.

The two cases presented by Pötzl and Fabbro provide us with an intriguing insight. These two individuals understand both languages very well. Hence, these two syndromes do not involve language loss. Rather it is a problem with the ability to access each language. Researchers have come to think of this as a problem with language switching.

The metaphor of a language switch, originally proposed by Wilder Penfield in the 1950s (Penfield & Roberts, 1959), has gained popularity in the literature in recent years. The idea is that bilinguals have to use general control mechanisms in order to access each of their two languages at the appropriate time. However, the importance of control has also been explored in bilinguals relative to monolinguals. These findings suggest certain advantages for bilinguals on tasks that, at least on the surface, have nothing to do with language.

The three themes that arise from the bilingual aphasia literature, AoA, proficiency, and cognitive control, serve as the main guiding posts in the present book. Hence, this book is divided into three main sections.

The first section, which consists of two chapters, expands on the role of AoA in determining the nature of the bilingual mind. Chapter 2 presents a model for understanding how language and mind develop. For centuries, philosophers, theoreticians, and researchers have debated two opposing viewpoints on how humans acquire knowledge—nature versus nurture. Is knowledge innate or learned? I consider the issue of language and cognitive development in light of this question. A review of the literature quickly leads us to realize that certain things are learned earlier than others. Also discussed is recent work that suggests a different metaphor for language learning. In this view, early language learning may actually build itself using basic sound elements of language. The ability to make fine phonological distinctions even for sounds from an unspoken language is present very early in life. In addition, young children spend hundreds of hours focusing on the spoken form of language. This emergentist approach in which larger skills—like language—are built from the bottom up provides the overarching themes that are explored further in the second chapter of this section.

In chapter 3, we expand this view to consider how age affects language learning in bilinguals. We begin by considering the nature of language learning in bilingual children. One of the central questions in the literature is whether bilinguals use one system or two. Initially, some researchers have suggested that bilingual children start off with a single integrated system that later develops into two. However, researchers have found the children can differentiate both languages very early in life. In addition, there appears to be considerable flexibility in the system. Skills appear to transfer fluidly from one language to another. Results from these studies add further support for the emergentist notion that was proposed in chapter 2. Specifically, bilingual children build up two language systems from the start. This creates a different neural representation than that seen in adult learners of a second language.

The second section of the book considers the role of proficiency in the neural and cognitive processing in bilinguals. This is covered in two separate chapters.

In chapter 4, we consider the nature of proficiency, starting with the wider issue of expertise. Expertise exists across many domains. This is best exemplified in the work of K. Anders Ericsson, who has found a set of general principles that apply across domains. Specifically, expertise is built up across time through deliberate practice in which individuals seek out or receive specific feedback about what is correct or not. This is thought to occur across many nonlinguistic domains. For example, recent work suggests that adults show similar types of acquisition profiles for nonfaces as those exhibited by children for faces (Gauthier & Nelson, 2001). This view is also mirrored in the monolingual literature, where adults and children are seen as utilizing similar skill acquisition systems (M. H. Johnson & Munakata, 2005). In its essence this view suggests that practice leads to changes in proficiency or expertise that are specific to language.

Chapter 5 considers the neural bases of language proficiency, focusing on bilinguals. Recent work on language proficiency has begun to identify the neural underpinning of improved skill. In general, experts tend to show smaller foci of activity relative to nonexperts. Most importantly, language proficiency seems to affect certain types of functions more than others. Language functions that involve the processing of meaning are more highly affected by language proficiency. Others, however, are less likely to be affected as strongly by language proficiency. For example, accent is much less likely to improve past a certain point even with continued increases in vocabulary and other higher-level forms of language in second language learners. Hence, proficiency has a bigger impact on certain language forms than on others.

The third section of the book moves beyond the AoA/proficiency factors that dictate the neural representation of language and considers the role of cognitive control in language processing. As noted earlier, researchers have used the metaphor of a language switch to explain a third type of deficit in which patients exhibit an inability to control which language is being spoken. The two chapters in this section shed light on current research in this area.

Chapter 6 addresses the nature of control in the context of monolinguals. I discuss how control develops across childhood and eventually declines in older adulthood. I also discuss how control plays a central role in task- or set-shifting. Chapter 7 expands on the nature of control within the bilingual context. Both Pitres and Ribot alluded to the fact that language loss could reflect temporary difficulties that were not directly due to damage to the language cortex. Pötzl took it one step further by suggesting that some regions of the brain were specialized for controlling access to one or the other language. The notion of a switching mechanism within the brain had already been proposed by Penfield, who suggested that bilinguals might employ a switch that could be used to maintain access to each language. In the 1960s and 1970s, a series of studies on language switching were conducted, with mixed results. The studies often explored whether there was a cost to switching and under what conditions this cost appeared. This increased cost could be in the form of absolute reaction time or in reduced ability to process the meaning of words in mixed language conditions. In the late 1990s the field turned to brain imaging techniques in order to uncover how the brain handles language switching. During the first decade of the 21st century, fMRI research led to one clear finding. The brain areas involved in language switching do not appear to be specific to language. Rather it appears that the bilingual brain uses areas involved in cognitive control when switching between languages. Finally, there is evidence that these control mechanisms may also get recruited in a less proficient second language. This suggests that bilinguals may be using more cognitive control in their everyday lives.

The importance of control also appears to have unintended benefits. Early research found that bilingual children showed better flexibility. For example, they understood that the sun (e.g., the object) was not the same as the word "sun" at an earlier age than monolingual children. This early ability to distinguish a word from its concept is due to the fact that bilinguals learn very early that there is more than one word for the same thing. Pioneering work by Ellen Bialystok found that this flexibility in the language system extended to tasks that require attentional flexibility. Furthermore, recent work suggests that bilinguals are able to modulate control on nonlinguistic attentional tasks. This bilingual advantage is most likely to appear in a person's 30s or 40s but is much more visible after age 60.

Chapter 8, the final chapter, brings these three strands together into a more coherent whole. In the first section, we considered the role of AoA in language learning. What emerges from the developmental literature is a clear view of how the brain is built up in expanding and overlapping systems. The earliest systems are devoted to sensorimotor processing. As development proceeds, later developing brain areas (e.g., the prefrontal cortex) become devoted to more complex processing. However, adults come to the language learning environment with a fully developed system. The use of these higher order systems provide certain advantages when learning explicit rules. However, this specialization naturally impedes the use of sensorimotor processing in later learned languages. In the second section, we considered the role of proficiency. Proficiency is interesting because it builds over time and is the result of intense practice. In the bilingual domain, proficiency has come to be associated with higher-level cognition, which appears in the processing of semantic violations and in certain higher-level grammatical forms. Finally, the third section builds on these earlier two sections by considering the role of cognitive control. There is clear evidence that the brain areas involved in language switching are involved in many other cognitive control tasks. Newer work suggests that bilinguals may show advantages when performing tasks that require cognitive control. This link is particularly interesting in the elderly, who exhibit larger language switching effects relative to young bilinguals but show significantly larger advantages relative to monolinguals on nonlinguistic control tasks. These results point to a link between cognitive control and language processing. In short, the neural bases of bilingualism are defined by AoA, language proficiency, and cognitive control. All three factors work together across time.

The book begins with a paradox. There are cases of patients who lose a language that seemingly lead us to think that bilinguals have separate areas for each language. The reality is much more complex than that. The reason for this complexity has to do with the nature of language itself. What researchers have come to call language is actually a collection of things that develop over a protracted period of time. This collection of things is affected by three different factors. AoA affects the order in which these things are put together. If two languages were assembled in slightly different ways, then they might actually break apart differently. In addition to this there is the component of expertise in a language that develops over time. This expertise is due to the amount of use over time. And finally, there are different control components that come into play that may affect one language and not the other. To date many researchers have suggested different types of models. Some have suggested an information processing approach, in which bilingualism is conceptualized as a set of mental programs. Others have suggested a more localizationalist approach, in which different languages are placed in specific areas of the brain. Others have taken a more linguistic approach, in which they describe the structure of each language and how it breaks apart after brain damage. These different approaches are all rooted in different disciplines and emphasize different aspects of language processing. Whereas each of these serves as a framework for certain aspects of the field, they have fallen short in attempting to provide a framework for understanding the neural bases of bilingualism.

What I propose in this book is to extend recent work that has begun to take a biological or natural systems approach. In short, I propose that two languages live inside one brain almost as two species live in an ecosystem. For the most part they peacefully coexist and often share resources. But they also compete for resources especially when under stress, as occurs when there is brain damage. Finally, the nature of bilingual language processing is intimately intertwined with the developmental profile of the brain. Because some brain areas develop earlier than others, this plays a role in the way in which early and late learners of a second language are instantiated in brain activity. That is, each language will take root in a network of brain areas depending on when that language is learned and at which rate it is learned. An interesting aside here is the relatively slow development of the frontal lobes, which coincides with its slower developing ability to perform cognitive control tasks. In a similar vein, learning languages at different ages should be affected by and affect the particular control circuits used by different individuals. Although there are still many questions to answer and many puzzles to solve, this book will argue that the nonlinear dynamic models that have been used to uncover the underlying mechanisms seen in natural systems and more recently in language and cognition, can be used to shed considerable light on the neural bases of bilingualism.

2

THE SENSORIMOTOR NATURE OF EARLY LEARNING

RIBOT, THE FOUNDER of French psychology, who came up with the law of regression, was neither the first nor the last to think that age plays a crucial role in how our minds are put together. His view, that "organic" types of memory are deeper and more engrained than others, has been taken up in modern literature. The influence of Ribot can be seen along two different lines. First, there is still considerable discussion about the ways that age of initial learning plays a role in the nature of adult processing. This is reflected in the classic question about whether nature or nurture plays the strongest role in the development of knowledge. The issue of nature versus nurture has also emerged in considering whether exposure to particular types of information during critical or sensitive periods leads to very particular patterns of acquisition. Furthermore, even the way in which we remember, read, or make decisions about words has been found to be constrained by age. This newer literature is based on the fact that certain types of cognitive and perceptual functions are constrained very tightly by the age of first exposure. Second, Ribot's influence can be seen in studies looking at how the brain changes from infancy to adulthood. The brain shows a pattern of development over time. The question that remains is which areas develop early in life and which develop later in life.

BRAIN, DEVELOPMENT, AND SENSORIMOTOR PROCESSING

The brain is at the most basic level a sensorimotor machine. It consists of four lobes, frontal, parietal, temporal, and occipital. Three of these lobes are devoted to sensory processing in some form. In the back of the brain lies the occipital lobe, which is devoted to visual processing. The temporal lobe, located a bit behind what are traditionally called the temples, is right beneath the ears and is known to be involved in the processing of sounds. The parietal lobe, in the upper middle portion of the brain, is highly interconnected with the other sensory areas of the brain and is thought to serve as a transfer point for the streams of information that bombard our senses. Hence, the entire back of our brain is devoted to sensory processing in some form.

Sensory input comes into the brain via a set of relays. It begins with the sensory signal arriving at the receptors. For the visual and auditory system, the receptors are located in the retina at the back of the eye or in the cochlear nucleus just behind the eardrum. These receptors convert light or vibrations into nerve impulses and transmit them to the corresponding nerve, then on to the thalamus, where they are relayed to sensory cortex. In the case of vision, sensory cortex is located in the occipital lobe. The auditory stream feeds into the temporal cortex. A similar stream exists for the touch sensations, which from nerves in the periphery shuttle up the spinal cord, through the ventroposterior thalamus, and into parietal cortex (Kaas, 2004).

Sensory maps spread out in similar fashion from their respective cortical centers. The temporal lobe has primary auditory areas with adjacent areas representing more and more complex processing of sound. The occipital lobe has primary visual areas, which then fan out and represent more and more complex forms of visual processing. From these base sensory areas information is broken up into different streams of processing (Gazzaniga et al., 2009). In the visual system, a lower stream that extends to the temporal lobe has been shown to be involved in the processing of objects. A superior stream fans out to the parietal cortex and has been shown to be involved with the location of things in the visual field. The breaking up of information into streams of processing has also been found in the temporal cortex (Hickok & Poeppel, 2004, 2007).

The bridge between the sensory and motor worlds is most dramatically represented at the border between the parietal and frontal lobes, the postcentral and precentral gyrus (see Figure 2.1). Here two body maps, mirror images of each other, clearly mark the brain's sensorimotor divide. The posterior parietal twin maintains a sensory map of the entire body, whereas its anterior frontal twin maintains the corresponding motor map. In these maps the body parts are not represented in proportion to their size but in proportion to the amount of stimulation or movement that a particular body part experiences. For example, the hands have much larger sensory

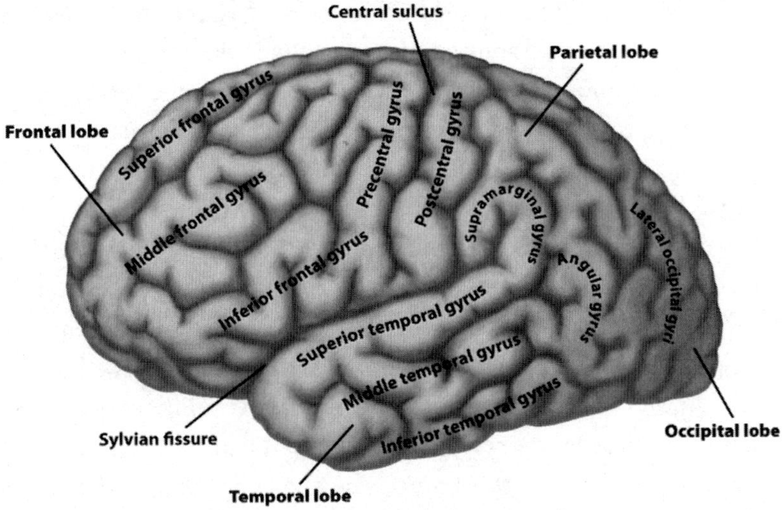

FIGURE 2.1 The main structures of the brain. Reprinted with permission from Gazzaniga, M. S., Ivry, R. B., & Mangun, G. R. (2009). *Cognitive neuroscience: The biology of the mind*. New York: W.W. Norton.

and motor maps than the feet. If you could visualize these maps as a person you would see a grotesque figure with huge hands and mouth but very small feet.

The importance of sensorimotor processing can also be seen in brain development. A number of neurological changes serve as markers of this development. Early in development, the brain produces many neurons and connections between these neurons, called dendrites. Over time many neurons die and the connections that exist between neurons are lost, a process called dendritic pruning. At the same time, the density of the myelin sheath that helps speed up the electrical signal sent by the axons increases (Campbell, 1986).

Interestingly, these markers of neural development do not spread evenly across the brain, at least not in humans. Unlike many other animal species, humans show a disjointed pattern of brain growth. Neurons mature earlier at central points and then fan out to areas further and further away. In an attempt to consolidate all available data on the rate at which different brain regions mature, Best (1988) proposed that growth occurs in three different ways: right-to-left, primary-to-secondary-to-tertiary,[1] and basal-to-cortical (from the middle of the brain out to the cortex).

More recent studies using a variety of techniques have confirmed the importance of these axes of brain development. In young infants, sensory cortices develop earliest in life. The development of sensory cortex is followed by development of sensory bridges in the parietal lobe and motor cortex in the frontal lobe. The most anterior regions of the brain in the prefrontal cortex are known to develop the latest. Although less dramatic than the brain changes seen during infancy and early childhood, changes in brain structure have been found to continue in middle childhood, across adolescence and adulthood, and through older adulthood. These findings have been confirmed by studies using direct observation of brain tissue as well as those using indirect methods of measuring the brain such as magnetic resonance imaging (MRI), which measures brain structure, and functional magnetic resonance imaging (fMRI), which looks at brain metabolism.

Neuronal changes that occur across development were first investigated by observing the anatomy of neurons in the brains of a few individuals under a microscope (Huttenlocher, 1990, 1994; Huttenlocher & Dabholkar, 1997; Huttenlocher & de Courten, 1987; Huttenlocher, De Courten, Garey, & Van der Loos, 1982a, 1982b). These initial studies fell along the lines discussed above. Differential development of areas under the microscope revealed changes in the production of synapses and in later synaptic pruning. The third process involves myelination, the addition of a fatty sheath surrounding the axon of each nerve cell. Production of synapses is strongest in the occipital lobe (i.e., visual regions) of the cortex between 4 and 8 months of age (Huttenlocher & de Courten, 1987). In the frontal lobe, synapse production reaches its peak at 15 months of age (Huttenlocher & Dabholkar, 1997). The lag in the frontal lobe's overproduction of synapses is also observed in the reduction of synapses via pruning (Huttenlocher, 1994).

The advent of newer neuroimaging techniques has allowed researchers to look at brain development in vivo. Work using these newer techniques has found evidence that developmental changes in brain areas can be seen as involving both a left-right, anterior-posterior gradient and a primary-to-secondary-to-tertiary gradient. Recent neuroimaging studies of single word processing with children have shown changes in the magnitude of neural activity in adults and children centered in areas of the frontal lobe (Schlaggar et al., 2002), the area that has been found to reach adult levels of synaptic connectivity the latest. This is consistent with the view that children have less developed frontal lobes relative to adults.

In a groundbreaking study Sowell et al. (2003) examined gray matter density using high resolution structural MRI scans in a group of individuals who ranged between 7 and 87 years of age. Results revealed a number of changes in density of gray matter, the part of the cortex composed of neural cell bodies that carry out neural computations. Across ages most areas showed a linear decrease in gray matter density. The authors speculate that changes in gray matter density may be due to increased myelination, which leads to increased white matter. For example, regions in the frontal lobe showed decreases in gray matter density from ages 7 and beyond. The authors attribute some of this to an increase in myelination up until age 40 with a decrease in neural density ensuing into older adulthood. An interesting deviation from this general trend was observed in the left temporal regions (in particular the posterior portions of the middle and superior temporal gyrus), which revealed an increase in gray matter density until age 40. These studies converge with previous work that had found continued changes in left temporal cortices well into adulthood. The late maturation of the temporal lobes particularly in the superior and posterior parts is an example of primary-to-secondary-to-tertiary development. In this particular case, it would involve extensions of auditory processing into language processing.

AGE OF ACQUISITION IN A SINGLE LANGUAGE

The development of early sensorimotor effects and later developing frontal lobe functions is a theme that can be seen in the word recognition literature. However, this detail came from an unexpected source. One day when I was a very green assistant professor, I headed to the library in search of an article on... Actually, I don't even remember. The only thing I know was that it was on experimental psychology. I found my article and began to scan the journal's table of contents. As I scanned the page, the title of another article caught my attention. The article was on age of acquisition (AoA) and word recognition. At first I thought it might be an article on bilingualism in which the authors looked at the effect of *second language* AoA. What I realized very quickly was that this article was actually on word recognition in monolinguals.

In this study, Catriona Morrison and Andrew Ellis (Morrison & Ellis, 1995) asked participants to respond to a set of visually presented stimuli by reading or identifying a visual stimulus as a word (i.e., "shave") or not (i.e., "mave"). In one set of experiments, they chose a set of early- ("fairy") and late-learned words ("wharf") that were matched for frequency, a measure of how often a person encounters the word. In a second set of experiments, they chose a set of high- ("market") and low-frequency ("pigeon") words that were matched on AoA. By doing this they could measure to what extent frequency of use or AoA influenced the speed with which a word was read or decisions were made. The results revealed effects of AoA but no effect of word frequency.

For most people this might seem like a sensible result, given Ribot's law of regression. Things learned earlier in life have stronger memories. However, at the time, these findings called into question the results of previous research that had focused extensively on word frequency. These studies had identified word frequency as the primary determinant of speed and accuracy in word recognition experiments (Besner & Smith, 1992; Grainger, 1990; Ostergaard, 1998). Morrison and Ellis showed a significant flaw in these previous studies that had failed to control for AoA. By identifying this new factor as a predictor of reading speed, this study showed that what researchers thought was a word frequency effect was most likely an AoA effect in disguise.

What caught my attention was the fact that the learning history of a word played a role in how fast monolingual adults could read or make decisions about it. A light bulb turned on in my head. Not only was AoA a determinant of differences across a bilingual's two languages but it was also a general property of words in a single language. With this fact in mind, I set out to look at whether we could extend this effect using a picture-naming paradigm. Gayane Meschyan, who was starting as a first-year graduate student, was in search of a master's thesis project. She and I discussed the research and she found it intriguing. But we both agreed that it would be nicer to actually use a 2 x 2 design, in which we could look at both AoA and word frequency independently. With this in mind, Gayane set out to design the experiment.

It took Gayane quite a bit of time to come up with the stimuli for this experiment. First of all, we had to choose from a set of pictures we knew would elicit low error rates. At the time, Elizabeth Bates, my doctoral advisor, had conducted what she termed "the mother of all picture naming studies," later known as picture norming (Bates, D'Amico, et al., 2003). Gayane was able to use Bates's database along with a separate study by Morrison (Morrison, Chappell, & Ellis, 1997) to arrive at a set of 18 pictures in each condition. Particularly challenging was the search for items that were learned early but relatively low in frequency ("pig") and those that were late and high in frequency ("hair"). Using pictures added a number of constraints. For example, we had to make sure that items across conditions were matched on visual complexity. This was done to eliminate the possibility that slower reaction times might be due to the fact that some pictures simply carried more information than others.

The experiment that Gayane conducted helped us to answer whether we could isolate an effect of both word AoA and word frequency (Meschyan & Hernandez, 2002). The answer was yes, at least for the pictures that we used in that experiment. Our findings showed that both AoA and word frequency had effects on reaction time. But we were left with a few questions about the difference between early- and late-learned words. What theory could account for this effect? Most importantly, would this theory shed light on the difference between first and second language learning?

In the summer of 2001, I traveled to Leipzig, Germany, to spend a month working with Sonja Kotz and Angela Frederici at the Max Planck Institute (MPI) for Cognitive Neuroscience. Aside from the incredible facilities and the state-of-the-art equipment, there was also a cadre of energetic young researchers who were busy working on the brain bases of language processing. Christian Fiebach, at the time a graduate student working at the MPI, had been interested in the nature of word frequency and its neural bases. As I looked at his data, it suddenly occurred to me that he had not considered whether his effect was due to word frequency or whether it was an AoA effect in disguise. So I asked him, "what about age of acquisition?" To which he responded, "age of acquisition?" I quickly sent him a few articles, and we met shortly thereafter to discuss them. We agreed that we should try and test whether there were independent effects of AoA and word frequency in his data.

Fiebach quickly collected some rated AoA measurements from subjects and we designed a study that would use monolingual German subjects. We decided that a regression approach would be best. Hence, we would statistically (rather than experimentally) isolate each factor. Christian's fear that AoA might "swallow up" his frequency effect was quickly allayed when he learned that the effect of frequency with AoA taken out was almost identical to the one in his published paper. What was most surprising was the effect of AoA (Fiebach, Friederici, Muller, von Cramon, & Hernandez, 2003). Words learned in early childhood led to activity in Heschl's gyrus and brain areas involved in the processing of speech sounds. Words learned in late childhood relied to a greater extent on brain areas in the lower portions of the inferior frontal lobe. These areas have been associated with the effortful access to a word's meaning (Badre, Poldrack, Pare-Blagoev, Insler, & Wagner, 2005; Bookheimer, 2002; Poldrack et al., 1999; Wagner, Maril, Bjork, & Schacter, 2001; Wagner, Pare-Blagoev, Clark, & Poldrack, 2001).

The divide between early- and late-learned words in the brain images bears an interesting resemblance to the way in which the brain develops. As I mentioned earlier, areas in sensory cortex are the earliest to develop, and those in the prefrontal cortex develop later. The results were fascinating because they suggested that even as adults we access words in the way that they were learned. Adults were "sounding out" early-learned words in their heads. For late-learned words, their brains acted as if they were working hard to make a link to other words with similar meanings.

Our findings also helped shed light on two theories that attempt to explain the single-word AoA effect in monolinguals. One prominent theory was that late-learned words are more phonologically complex than early-learned words. The idea is that early-learned words are stored as single sound units whereas late-learned words are built up in parts. Unfortunately, this theory has not received much support in the experimental psychology literature (Monaghan & Ellis, 2002).

A second hypothesis at the time held that early-learned words had the benefit of arriving first into the system. Think about a person who goes to a party full of strangers. If he or she arrives early, the person is likely to meet someone they might not have met before or to spend more time talking to the organizers. This person is more likely to experience and change the social dynamic of the party. The social fabric of the party will be driven by all of the people who arrive early. A person who arrives late is likely to come at a time when the social dynamics are already set. The late arrivers must integrate themselves into an existing structure.

Ellis and Lambon Ralph had a similar way of describing AoA effects based on a series of neural network simulations (A. W. Ellis & Lambon Ralph, 2000). Because early-learned words formed the basis of the system they would come to play a more central role in word-level representations. Hence, late-learned words would be built around early-learned words. In this case, it did not have anything to do with the phonology of the word. It had to do with the fact that early-learned words enter the lexical system first. As the structure gets built there is less flexibility for late-learned words to influence the system.

The findings from Fiebach's study shed interesting light on this issue. We did find evidence that early-learned words involved neural activity in areas associated with auditory processing. However, we also found evidence that late-learned words involved neural activity in brain areas involved in processing meaning. In some ways, these findings lend support to both theories of single-word AoA. Phonological aspects of a word are present during early learning. Later learning involves processing words using meaning.

AOA AND SENSITIVE PERIODS

The rapid maturation of sensory and motor cortices has also been tied to the long-standing literature that has discussed the nature of critical periods. Evidence of AoA effects has been found in many nonlinguistic domains. For example, it is well known that early deprivation or alteration of sensory input leads to impaired sensory perception in many species. This is known from the pioneering work of Hubel and Wiesel, two neurophysiologists who were interested in uncovering how the visual cortex develops. Their initial work was designed to try and make a link between single cells firing in the cortex and visual stimulation.

When Hubel and Wiesel began their search, they were building on the work of Kuffler, who had established the center-surround properties of ganglion cells in the cat retina (Kuffler, 1953). So they placed slides on an overhead projector a few feet in front of the cat to see what might make cells in the visual areas located within the occipital cortex respond. Initially, they used slides with dots, as Kuffler had. They searched for weeks with no response. They also tried other visual stimuli on the projector. Time after time there was no response, no firing, just silence. Then one day out of the blue a cell fired. Hubel and Wiesel wondered what might have led to the firing. There was nothing different about the particular slide they were using. After some time they realized that the firing was not due to what was on the slide. What happened was that the slide was misaligned, thus creating an edge of light and darkness. Cells were reacting to these edges and not to anything on the slide. Serendipity led them to the initial discovery that would later serve as a foundation for many of the future findings in visual cortical neurophysiology.

Hubel and Wiesel continued their work in this domain by looking at visual critical periods. In a series of papers, they outlined how visual deprivation affected the neural structure of the lateral geniculate nucleus in the thalamus and the primary visual cortex. They found that young kittens had a sensitive period during which deprivation of visual input would lead to long-term visual impairments (Hubel & Wiesel, 1963; Wiesel & Hubel, 1963a, 1963b). Interruption of visual input in one eye, however, did not impair vision in adult cats. Hence, Hubel and Wiesel showed that exposure to visual stimuli early in life has profound effects on primary visual cortex.

Since Hubel and Wiesel's groundbreaking findings, researchers in the field have found evidence of a sensitive period in cats, monkeys, rats, mice, ferrets, and humans (Banks, Aslin, & Letson, 1975; Fagiolini, Pizzorusso, Berardi, Domenici, & Maffei, 1994; Harwerth, Smith, Duncan, Crawford, & von Noorden, 1986; Huang et al., 1999; Issa, Trachtenberg, Chapman, Zahs, & Stryker, 1999; Olson & Freeman, 1980). Similar research also reveals critical period effects in the calibration of the auditory map by visual input (M. Brainard & Knudsen, 1998).

Sensory deprivation can also lead to problems in the motor system. For example, disruption of binocular experience can lead to problems that adversely affect smooth pursuit of moving objects and can lead to eyes drifting when viewing stationary targets (Norcia, 1996). Hence, problems in the sensory domain lead to abnormalities of motor function.

The effects of AoA have also been seen in learning of birdsong, which has been conceptualized as involving three phases, sensory, sensorimotor, and crystallized (M. S. Brainard & Doupe, 2002). During the sensory period, a bird listens to the song of a tutor in order to form a template. Lack of exposure to an adult song during this phase leads to irregular songs that contain some species-specific characteristics. In the second sensorimotor phase, the bird learns to match the song to the template. During this phase songbirds fine-tune their songs through practice. Auditory feedback is crucial during this time. In the final crystallized phase, birds are mature and can produce the species-specific song but may no longer be able to learn other songs as well. In short, learning of birdsong also points to a sensorimotor basis for this early form of learning followed by a longer period in which the bird is no longer able to produce different types of songs.

Finally, differences in AoA have also been observed in higher-level nonlinguistic functions. Work with musicians has found effects of AoA at both the behavioral and neural level. There is evidence that absolute pitch can be learned by speakers of nontonal languages only before the age of 7 (Deutsch, Henthorn, Marvin, & Xu, 2006; Trainor, 2005). In addition, the ability to synchronize motor responses to a visually presented flashing square has been found to differ significantly between groups of professional musicians as a function of AoA, even when these groups are matched for years of musical experience, years of formal training, and hours of current practice (Watanabe, Savion-Lemieux, & Penhune, 2007). Work using fMRI has found that there is evidence that early musical training correlates with the size of the digit representation in motor regions of the cortex (Elbert, Pantev, Wienbruch, Rockstroh, & Taub, 1995). In a similar vein, Schlaug et al. (1995) found the anterior corpus callosum (one of two tracts that connects the right and left hemispheres) to be larger in musicians than nonmusicians, and largest for those who learned to play before the age of seven. Hence, the effects of AoA on behavior as well as neural representations in the music domain can be viewed as reflecting sensorimotor processing.

The fact that sensorimotor processing might be playing a role also led me to think about a domain that captures the fascination of people across the globe. Specifically, I wondered whether sports might also show similar types of effects. The notion of expertise is one that has been discussed to a great extent across many domains (for a review see Ericsson, Charness, Feltovich, & Hoffman, 2006) and as it relates to motor skills in athletes (R. Gray, 2004; S. Gray, Watts, Debicki, & Hore, 2006). Of particular interest was work done by Sian Beilock comparing novices and experts. In one experiment published by her group, expert golf players were asked to perform a simple putting task under one of three conditions: putting freely, putting while listening to a set of words, and putting while paying attention to a particular body part. The latter two were both dual task situations in that they required participants to putt and do something else. In one condition, participants were asked to say out loud some fact about a body part during or after training. In the other condition, they were asked to tell an experimenter all the words they remembered during the putting task.

Beilock found that performance in experts dropped when they were attending to a particular body part relative to the condition where they had to remember words (Beilock, Carr, MacMahon, & Starkes, 2002). Interestingly, novices showed the opposite effect. They were better when attending to a body part but relatively worse when they had to remember words. Beilock suggests that experts and novices differ in the extent to which they use procedural memory, a form of memory that becomes automatized in the performance of motor skills.

Reading through these studies, it was somewhat surprising that no one had really considered the role of AoA. As in the AoA literature on words, it seemed likely that expertise might have served as a proxy for the age of initial learning. Beilock and I discussed it and decided that it would be an interesting line to pursue.

A few months later, Beilock and I were corresponding over e-mail when she told me about an idea she had. She thought the same logic that helped to distinguish between novices and experts might serve as a behavioral test of the sensorimotor AoA hypothesis. Specifically, she thought that early learners might have more trouble putting a golf ball when attending to their physical movement. Late learners, on the other hand, would have more trouble putting when listening to words. Her idea was that early and late learners would look like experts and novices, respectively.

Richard Redding, a master's student at the time, found two groups of golf players who varied in the age at which they had learned to play golf. The early group had learned to play golf before the age of 10 and the late group after the age of 10. The results revealed a very interesting split between the early and late learners. Those who learned to play golf before the age of 10 showed significantly poorer performance when asked to monitor their body's movement relative to late learners. There was also a trend for the late learners to have more difficulty recalling words in the dual-task condition. Finally, the two groups were indistinguishable when putting without a task.

The results from this study with athletes suggests that skilled motor processing shares some similarities with other sensorimotor skills such as learning words and with learning in sensory domains such as vision. Specifically, early learners seem to use a more embodied representation to perform golf putting. You could say they "feel" the putt more. Having to actually break it up into parts and report some piece of it hinders their performance. For late learners, the putt is done in a more conscious manner, and hence they have access to conscious information which they can then report verbally. As in other domains, adults who learn a sport as a child appear to adopt a stronger sensorimotor representation relative to those who learn it at an older age.

Given these results, Beilock and I felt that it would make sense to peer inside the athletes' brains to see whether differences in AoA revealed differences in the brain representations associated with their sport. But neither of us had funding for this study. Our chance to collect this data came by happenstance.

In 2009 the University of Houston (UH) had entered into an agreement with the Human Neuroimaging Laboratory (HNL) to pay a single sum for all scanning during the year. This meant that we could scan subjects without paying for each one of them. Read Montague, the HNL's director, reasoned that the majority of the cost of running a scanner had to do with maintenance. It actually did not cost that much money to use it more often. Hence, he felt that getting institutions to pay up front would generate more revenue. Nickel and diming people for each scan would actually lead people to spend less rather than more.

In addition to providing a means to cover maintenance costs, the agreement between UH and the HNL came with a significant scientific payoff. Read felt that a major impediment in fMRI research was the feeling that people were stuck with too little power to test the most interesting questions. The UH-Baylor scanning agreement would allow people to run as many subjects as they needed to complete a study without worrying about saving money on a grant. Read's foresight allowed us to run many more subjects than we had before. It also allowed us to run studies that we would not have even dreamed of running. Our work with athletes, which was not part of my regular funding, is an example of this. And so we set out to test a group of athletes in an fMRI task.

The one issue with our behavioral study was that it involved an overt response with a clear manipulation. But I felt that the sensorimotor effects of early learning would most likely reveal themselves in a more subtle manner. If, in fact, taking a sensorimotor chunk and breaking it apart in a lab hurts early learners more, then we should expect a tighter coupling between a particular sport stimulus and its response. To test this idea we went out and recruited a group of college and recreational tennis and basketball players. During the first part of the experiment they were asked to simply listen to a set of sounds that were either from their sport (i.e., a tennis player hearing tennis sounds) or the other sport (i.e., a tennis player hearing basketball sounds). We also presented them a series of familiar (phone ringing) and unfamiliar environmental sounds. Because each sound was so short we presented it five times in succession, in order to make sure that participants could hear each sound.

Because we were testing a diverse group of tennis and basketball players, there was considerable variation in the age at which a particular individual began to play their preferred sport, which ranged from 2 to 12. This allowed us to look at how AoA might influence the brain signals associated with hearing tennis or basketball sounds. The results were stunning in their simplicity. Early learners showed increased activity in the right and left hippocampus. Late learners revealed activity in the right and left superior frontal gyrus, the inferior parietal cortex, and the precuneus. These results revealed an interesting split between the neural circuits that are involved in the recognition of sounds for the most familiar sport. Specifically, early learners actually began to form memory traces for the sport they played. Late learners, on the other hand, activated a neural circuit that involved a multisensory spatial representation coupled with areas that have been found to be related to representation of the self. It is almost as if the early learners were storing the sounds for future reference. In late learners, familiar sport sounds were activating a more explicit memory circuit. Hence, the age at which a sport is first learned dictates the type of memory circuit that is invoked by sounds in the most familiar sport.

When Beth Owens Woods and I were discussing the experiment in the early phases, I was really worried that we would not find anything. I had never used passive listening in the scanner and I really worried about the possibility that we would have no significant differences across our conditions. I insisted that we also have a more demanding task given to the same subjects. To accomplish this we adapted a task that had been used by Ken Kishida and Read Montague of the HNL. Kishida and Montague asked a group of college and professional athletes to watch video sports scenes. After watching each scene, they were asked to imagine seeing the video again (watch-it) or to imagine themselves performing the action seen in the video (do-it). Their results revealed greater activity in the do-it condition relative to the watch-it condition. This effect was particularly striking in the do-it condition, where athletes showed a lot of activity below the surface of the cortex for their own sport. When trying to imagine the motions for the other sport they showed much less activity.

It occurred to us that we could use a similar type of manipulation with our athletes. Specifically, we asked the athletes to listen to sounds and then instructed them to imagine hearing it again (hear-it) or to imagine doing the action (do-it). Interestingly, the differences between these two conditions were not as great as those observed by Kishida and Montague, which suggests that converting a sound into an action may not be the same as converting a series of images into an action. This led us to combine both the hear-it and do-it conditions into a single condition. Unlike our passive listening condition, this combined condition involved explicit or conscious retrieval.

In order to uncover the effect of AoA, we performed a regression on the explicit condition using the same procedure we had used in the passive listening condition. This regression revealed a different type of split between early and late learners. Early learners showed brain activity in right and left hemisphere superior parietal cortex and in a second more anterior portion of the parietal cortex in the left rolandic operculum. Late learners showed activity in a more superior portion of the left parietal cortex, the left motor cortex, and left Broca's area. There are a couple of interesting things to notice here. First, the early learners show relatively more brain activity in the posterior parts of the brain. The parietal cortex is known to be involved in visual-spatial processing (Carpenter, Just, Keller, Eddy, & Thulborn, 1999; Jonides, Smith, Koeppe, & Awh, 1993; Rowe & Passingham, 2001). The presence of both right and left hemisphere activity suggests that early learners are more likely to invoke a spatial representation that is present across their entire bodies. Hence, early learners seem to be activating a representation that exists in space to a much greater extent than late learners. Late learners, on the other hand, show increased activity in a more distributed motor-skill circuit. This circuit involves connections between superior parietal cortex and two areas in the frontal cortex, one involved in invoking the appropriate motor areas and a second involved in a motor plan. In short, early learners show a more spatial representation, whereas late learners show a more controlled motor-skill representation.

These results shed considerable light on our findings with early and late learners of golf. If you recall from that study, we observed that early learners had more difficulty unpacking their golf putt relative to late learners. We speculated that this might be due to the fact that early learners were learning their sport in a more sensorimotor manner. The results from our imaging study with athletes suggest that early learners seem to integrate their sports sounds with very specific purposes. When asked to passively listen to these sounds, they automatically store them for future reference. When asked to retrieve an explicit memory associated with these sounds, they invoke a spatial representation, almost as if their brains already had a plan in place for sounds in their familiar sport. Late learners, on the other hand, had to consciously pull up these sounds even when we asked them to simply listen. This explains why it was easier for them to tell us what they were doing. They were more aware of how these movements had been executed and hence had more access to a conscious representation of these memories. Our results suggest that early learners simply do things with little conscious awareness, late learners are more likely to be aware of what they are doing.

Based on a review of the literature so far, research shows an interesting link across a number of domains. AoA effects have been most clearly found in the sensory and motor domains. Furthermore, higher-level skills such as reading words, matching the rhythm of a flashing square, and putting in golf are influenced by the time at which skills are learned. Taken together, these findings suggest that early learning is more directly tied to bodily experience than late learning. Late learning seems to be more strategic and conscious. These results are also in line with the way in which cortex develops in that early learning is more closely tied to activity in more embodied systems whereas later learned skills appear to involve more complex networks that typically involve the prefrontal cortex. As we will see in the next section, moving from sensorimotor to cognitive processing serves as an interesting way to think about language learning in general.

AOA AND LANGUAGE

The notion of critical periods in development has also played a role in language. Work with single language learners finds clear evidence that age of initial learning affects the way in which language is learned. This starts with the age-old issue of the sensitive period for language learning. There is, of course, a large literature on the nature of critical periods in second language learning that will be covered in the next chapter. Here we will focus on evidence of critical periods from studies with single language learners.

Among the most contentious sets of evidence in favor of a sensitive period for language are the observations made with feral children. The two most famous cases are of Victor of Aveyron and Genie (Candland, 1993; Curtiss, Fromkin, Krashen,

Rigler, & Rigler, 1974). Victor was found in the forest just outside the town of Saint Sernin sur Rance in France at the end of the 18th century. He was taken in by Jean-Marc Gaspard Itard, who attempted and failed to make Victor into a "normal" adult. The story of Genie had a different beginning. Her father had professed that she was incurably mentally retarded because of a hip deformation that did not allow her to walk at a normal age. From the age of 20 months, she was strapped to a toilet seat and raised in almost complete social deprivation. Both her brother and mother were only allowed to bark at her. In addition, her mother was allowed to feed her and care for her only minimally.

On November 4, 1970, Genie was admitted to Children's Hospital in Los Angeles. Besides the many physical effects of the harsh environment in which she was raised, the most striking effect was her inability to communicate with others using language. Although her ability to comprehend words was present from the outset, her production lagged considerably. At first, articulation was particularly difficult and required considerable effort on her part. However, over time she began to produce language. In fact, her ability to produce language paralleled many of the stages seen commonly in children. For example, she began with simple consonant vowel combinations before transitioning to using more complex sound patterns such as liquids (the sound l in liquid) or glides (the w in wave). Although her speech was not completely normal, it did appear that she was able to produce all of the speech sounds present in the English language.

The similarities and differences between Genie and normally developing children were also apparent in her vocabulary and grammatical development. One interesting finding was that her vocabulary development seemed to be ahead of that seen in normally developing children. For example, at the typical two-word stage she had a working vocabulary of 200 words, whereas most children would be at the 50-word stage. Genie did show the acquisition of the basic word endings that signal grammatical category (i.e., jump, jumping). However, her ability to use grammar never reached that of an adult. Susan Curtiss, the researcher who worked with Genie most extensively, interviewed her for the last time in March of 1978. Genie was placed back into the custody of her mother and never again participated in scientific research.

The results from Genie suggest an interesting dissociation between the different aspects of language that are affected by early exposure. Specifically, it appeared that Genie had considerable problems with learning grammar and especially with grammatical endings. However, she was able to learn vocabulary much more easily. Finally, it appears that Genie also had difficulty with some aspects of language production. The fact that early language deprivation seems to affect grammar and phonology but not vocabulary opens up an interesting question. Specifically, it suggests that there might be a connection between speech sounds and the ways in which words are assembled into grammatically correct sentences. We will trace out the link that researchers have made between speech and grammar. But first let's discuss how infants learn to turn sounds into words.

DIGGING INTO THE DEPTHS OF LANGUAGE

Understanding speech for adults is so straightforward that it seems trivial. To understand a little better how difficult this task is it helps to actually see what speech is like. An example of this can be seen in Figure 2.2. At the bottom you can see the words typed out. The top portion shows you the spectrogram, where air movements are measured. The interesting thing is that there are no real word boundaries. I first noticed this when we were trying to splice out words from a sentence for an experiment. A research assistant went off with the task of cutting out certain words from a recorded audio sample of a sentence only to come back two minutes later unable to complete the task.

If you look at the figure you will notice very quickly that the phonemes (basic sounds like /n/ and /r/) are scrunched together. If you ever tried to learn a foreign language you will have experienced this first hand. Speech is so fast that it is simply a blur of sounds. Given that the world is buzzing with all these scrunched up sounds, one is left wondering how children can possibly make sense of it all.

The answer to this question has appeared in a somewhat surprising set of findings that have emerged from the developmental psychology literature. In a seminal set of studies, Jenny Saffran and colleagues wondered whether 8-month-old infants could learn to distinguish word boundaries (Saffran, 2001; Saffran, Aslin, & Newport, 1996). To test this, Saffran and colleagues presented children with a sequence of sounds during a familiarization phase. These consisted of a set of continuous sound patterns in which the probabilities of sound occurring next to each other were manipulated. The probabilities of sounds occurring next to each other (transition probabilities) are what distinguish word boundaries. Sounds that occur between different words may have low transition probabilities—that is, they occur together less frequently. For example, the *r* and *b* between "occur" and "between" in the previous

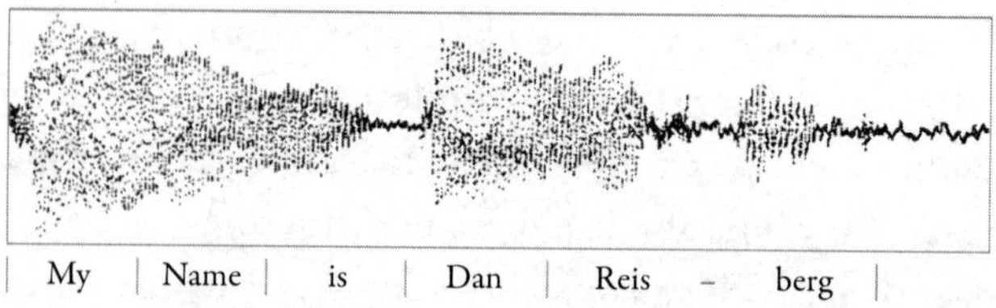

FIGURE 2.2 Typed out (bottom) and spectrogram (top) of the sentence "My name is Dan Reisberg." Reprinted with permission from Reisberg, D. (2010). *Cognition: Exploring the science of the mind.* New York: W.W. Norton.

sentence are much less likely to occur together than the *b* and *e* that start the word "between." If one simply kept track of the frequency with which these sounds occurred together one would quickly start to come out with a schematic for detecting word boundaries. The basic guide would suggest that sounds that occur together frequently appear in the middle of a word. Those that occur much less often mark the boundaries between words.

The use of transition probabilities and the places where they may be deceiving can be found in children's speech. For example, Kamille, my youngest daughter, always heard us tell her to be nice, strong, and careful. One day she turned around to my wife and asked her, "Mom am I being haive?" She thought that "behave" was two words not one. Adult English speakers would pick up that "beHAVE" is one word. Like the babies that Saffran tested, Kamille was using very simple sound statistics to estimate word boundaries. She had heard the sound "be" used with a bunch of different adjectives and assumed that "haive" must be an adjective. Since the verb "be" is so common, she interpreted "behave" as two words not one. In short, she was using the sound patterns of English to determine where to break up words.[2]

To hedge their bets, Saffran and colleagues used a transition probability of 100% to distinguish word boundaries. Eight-month-old infants were presented with these sets of sounds for 3 minutes. Afterward, they were tested using a head-turning procedure, in which head turning would indicate novelty. Saffran and colleagues found that infants turned their heads longer when listening to partial words than to whole words. This indicated that infants could learn the new word boundaries within a few minutes. This type of learning has been called statistical learning due to the fact that it involves probabilities as a method for deriving word boundaries.

Of particular interest to us is the fact that children can use chunking to meld sound pieces into single words. They do this with passive listening and a few tricks. The fact that children can learn in this manner suggests that they are expert learners who can very easily make sense of the linguistic chatter that occurs in their world.

Words are great things, and knowing where one word ends and another begins is important. But children also have to learn how to put words together depending on grammatical function. As discussed above, Genie was known for having particular difficulty with grammatical processing. Hence, the ability to use small changes at the word level to track grammar seems to be particularly tied to AoA.

BUILDING A GRAMMAR

The fact that speech sounds can help with uncovering grammar can be seen in the error made by my daughter Kamille. Whereas she thought that "be haive" was two words, adult English speakers would pick up that "beHAVE" is one word. Hence, knowing the probability that two sounds go together helps to distinguish words. However, it is not enough by itself to uncover the workings of grammar. The ways in which English speakers come to distinguish nouns and verbs illustrates this nicely. In English, the majority of two-syllable words contain the stress on the first syllable (e.g., trochaic). In addition, there are also a number of words that contain stress on the second syllable (e.g. iambic). However, these differences are not equally distributed across all word classes. Nouns mostly exhibit a trochaic stress pattern (PERson), whereas verbs tend to exhibit an iambic stress pattern (purSUE). Interestingly, noun-verb homographs can differ in their stress patterns such that they take on a trochaic pattern for the noun (PERmit) and an iambic pattern for the verb (perMIT). Hence, the auditory signal not only helps to distinguish where words begin and end but also allows us to distinguish between words that have different grammatical functions. The differences in stress patterns illustrate how grammatical categories can be signaled in the underlying sound structures of language.

Before proceeding with this discussion, let's stop for a moment and consider phonological development as an early marker for language acquisition. The very early phonological abilities of young language learners have been uncovered in the extraordinary work of Patricia Kuhl and colleagues (Kuhl, 2004). In a set of very influential studies, Kuhl and colleagues have found that young infants are born with the ability to detect subtle differences in phonemes in both their native and nonnative languages. To take the most classic example, a Japanese infant can detect the difference between the "l" sound and the "r" sound. This ability remains until about 6 months, at which time it begins to diminish. By the end of the first year, an infant exposed exclusively to Japanese can no longer distinguish between "l" and "r." These findings suggest that the detection of speech sounds becomes specialized in a native language during the first year of life (for further discussion see Werker & Tees, 2005).

Researchers have established two sets of clear and consistent findings. First, it is clear that speech recognition specializes within the first year of life and that the ability to recognize sounds helps in estimating word boundaries. Second, there appear to be clear differences in the sound patterns of words from different word classes. Earlier we noted that in English, two-syllable verbs and nouns place stress differentially on the first or second syllable. Thus at least for these two types of word categories, there appears to be a connection between sound and grammar.

It is precisely this view that has come to the fore of developmental psycholinguistics. The hypothesis that grammar learning may have its basis in the phonological aspects of language is more fully explored by Morgan and Demuth in their aptly titled book *From Signal to Syntax: Bootstrapping from Speech to Grammar in Early Acquisition*. In this book, the authors propose that the speech stream in combination with a child's limitations allow them to build up a full grammatical representation (Morgan & Demuth, 1996a, 1996b).

The ways in which grammar depends on speech sounds can be found in the types of errors that young children make in spoken language. For example, Gerken and colleagues (Gerken, 1994; Gerken & McIntosh, 1993) found many more instances of full noun phrases in the direct object position than in the subject position. That is, children were more likely to say things like "Doggie bites the book," but were much less likely to say things like "THE doggie bites book." But even the object determiners suffer from omission at times. For example, children are more likely to say "Dog bites the book" but not "Dog chases THE book." At first glance this might seem random. Demuth suggests that there are important prosodic differences between these two sentences. Namely, "Dog bites the book" blends together more easily than "Dog chases the book." If you trying saying both phrases you might notice that you have to pause after "chases" but not "bites" before saying "the book." The first sentence just rolls of your tongue much more easily. Demuth followed this up by observing children between 16 and 28 months. She found that children were much more consistent at producing the determiner "the" in the object position when the sounds blended together well (Demuth, Patrolia, Song, & Masapollo, 2012).

What about other languages? Well, interestingly, Spanish-speaking children tend to hear three-syllable words much more often. Hence, their auditory window is a little bigger. The larger auditory window allows the determiner (*el*, *la*) to be produced with much more consistency than in English. Hence, from the very beginning children learn determiner and noun together and produce them that way as well. Eventually, this leads the child into the complex world of grammatical gender markings which nonnative Spanish speakers find so difficult. However, there is no rule about gender marking built into the child. In Demuth's view, the links between a determiner (*el*/*la*) and the word happens because of the larger window that is used by Spanish-speaking children to parse words. Hence, children learn "elperro" (the-dog) as one word not two and they learn it a lot earlier. In short, gender emerges from a system that is setup to learn trisyllabic words. These prosodic differences are like larger melodic filters that allow speakers of a language to properly hear the most crucial pieces.

THE SENSORIMOTOR NATURE OF AOA

As we have seen in this chapter, age plays a role in the way in which individuals learn. AoA of single words has been shown to rely on different neural substrates. Whereas processing of early-learned words lead to activity in areas of the brain devoted to speech and sound processing, making decisions about late-learned words leads to increased activity in areas of the brain involved in the processing of meaning. Across the rest of the chapter, I reviewed a number of examples that show how sensorimotor functioning is tied to early perceptual and cognitive functions. I also reviewed examples of this within the domain of language and presented some evidence that processing of the speech signal serves as a basis for forming words and grammatical categories. The importance of sound in language can be seen in the fact that the left temporal lobe shows continued maturation into a person's 40s. Taken together, these results suggest that across development, higher-level categories are being built up from sensorimotor building blocks.

The particular view that I propose is not deterministic in the way that traditional sensitive period views of language are (Lenneberg, 1967). The fact that AoA plays a role in the processing of information is driven by a variety of factors, which include both external and internal sources. These sources lead to an uneven type of development, a pattern of brain growth that is particular to humans. This uneven pattern of development suggests that children and adults come to the learning of skills equipped with very different neural and cognitive configurations. It is the interaction between age and learning of a second language that is the main topic of the next chapter.

NOTES

1. An example of this can be seen in the visual cortex. Primary visual cortex represents visual space in a one-to-one fashion such that each point in the brain corresponds to a location in an individual's visual space. Secondary cortices re-represent these areas using more complex information such as motion or color. Tertiary areas may represent, for example, single objects in space which do not correspond to what is actually seen in particular locations in space. A similar phenomenon occurs in auditory cortex, where primary auditory cortex is involved in more primitive forms of auditory processing whereas secondary and tertiary auditory areas may represent other aspects of the auditory signal such as speech sounds, words, or grammatical aspects of the signal.
2. A later occurring error led Kamille to say that the bad guy was being "underarrested."

3

LAYING DOWN TWO LANGUAGES ACROSS TIME

IN THE LAST chapter we considered how reliance on sensorimotor processing changes as a function of age of acquisition (AoA). Results from a number of studies have found that as brain maturation expands from primary to tertiary areas within sensorimotor cortex and extends into the frontal lobe individuals begin to process information differently. The neurological changes that occur are also accompanied by changes in the concomitant behavior. This is captured in the way that researchers have conceptualized the learning of grammar as being based on the sounds of a particular language. It is also seen in the fact that sensitive periods during early development are observed mostly in basic sensory and motor functions. In the case of language, it involves multiple levels of processing that build on each other. Hence, certain aspects of language, such as grammar and phonology are more sensitive to deprivation, as we saw in the case of Genie.

The fact that different aspects of language are differentially sensitive to age has also appeared in the second language learning literature. Second language learners often struggle to master the phonology and grammar of a new language. Furthermore, this effect changes as a function of age. In this chapter, we will consider to what extent the effects of AoA on second language learning can also be conceptualized as involving sensorimotor processing.

LET THERE BE SOUND

As we noted in the last chapter, the building blocks of language are established at an extremely early age. Recent work from Silke Telkemeyer has found that the brain activity for native-language sounds differs from that for sounds from other languages within a few days of birth (Telkemeyer et al., 2009), a finding that is in line with earlier behavioral studies. Just four days after birth French babies suck harder when hearing French than Russian (Mehler, Jusczyk, Lambertz, & Halsted, 1988). After just a month of life, infants can distinguish between the sounds "ba" and "pa" (Eimas, Siqueland, Jusczyk, & Vigorito, 1971).

We have also considered two different aspects of speech. When you hear someone speak you hear the sounds within each word (linguists call them phonemes). This is probably most noticeable to you when you hear a foreign or regional accent. One time my doctoral advisor, Liz Bates, came back from Florida and recounted a bewildering conversation she had had with a local person. The person was telling her about this horrible rice problem they had in Florida. Liz was confused. She wondered if there was famine or if crops had been ruined by torrential rains. Well, it turns out that it had nothing to do with food. The person was telling Liz about the horrible RACE problem. The pronunciation of *a* in that regional accent was different enough to have thrown Liz off for just a few seconds. These within-word differences between an *a* (as in kate) and an *i* (as in kite) would be referred to as a difference at the phonological level.

In addition, to differences in the sounds within words there are also differences across phrases and sentences. Linguists call the latter melodic information in language prosody. The surfer dude at the beach and a young woman from the San Fernando Valley near Los Angeles don't just differ in the way they pronounce each word. They also differ in how a whole phrase of their speech sounds. The Valley girl accent that has spread across much of the United States is most notable for its use of the word "like" and the upward intonation you hear at the end of every phrase. This accent differs from its cousin the surfer dude accent, which tends to slow speech down and maintain a flat intonation which sounds more relaxed. Hence, Valley girls and surfer dudes differ in the melodic quality of language but not in the sounds within words. Technically speaking their English differs in prosody but not in phonology.

It is this intonation of language that drives infants' ability to distinguish speech. As noted earlier, experiments have revealed that all of us in our first months of life quickly learn to distinguish prosody in our language from that in another language. Furthermore, infants exposed to speech stripped out of detailed information (think of hummed speech) show the same effect. This suggests that infants are using larger melodic contours to differentiate speech in different languages.

Prosody, however, cannot be the only factor playing a role in language learning. Languages with similar rhythmic properties such as Dutch and English sound very different. Hence, Dutch- or English-speaking children show preferences to their native language at a very early age. One-year-old English-speaking children don't respond to Dutch adults. To achieve this feat, monolingual children have to learn during the first year of life to breach the rhythmic information present in the surface prosody across phrases and sentences in order to use the finer detailed information in the phonology present inside individual words. The case of bilingualism poses an interesting challenge to the developing cognitive system. The question is whether bilingual children are able to use fine-grained phonological information to distinguish two languages at the same rate as monolingual children.

Work by Laura Bosch and Nuria Sebastian-Galles (2001) explores the use of finer sound distinctions by comparing Spanish speakers to Catalan-Spanish bilinguals. Catalan is a romance language spoken in the northeast portions of Spain and in southern France. To the nonnative ear, Catalan would sound very similar to Spanish because it has a similar rhythm and because it has many words that share meaning and sound. However, Spanish speakers would struggle to distinguish certain Catalan vowels, which don't exist in their native language. Hence Catalan and Spanish differ in the phonemes (speech sounds) used within words but are very similar in their use of prosody (linguistic melody) across phrases and sentences.

Using these two languages Bosch and Sebastian-Galles sought answers to two questions. First, they wanted to know at which age monolingual Spanish and Catalan speakers could learn to differentiate these rhythmically similar languages. Second, they wanted to test whether Catalan-Spanish bilinguals would be slower to pick up these distinctions. In short, the question is whether bilingual infants might experience a period of brief "confusion" as they worked out the subtle differences between the sounds in each language.

To test phonological sensitivity Bosch and Nuria Sebastian-Galles (2001) asked 4-month-old infant bilinguals and monolinguals to listen to Spanish and Catalan during 30-second passages spoken by a monolingual female. Surprisingly, bilingual infants could distinguish between two rhythmically similar native languages like Catalan and Spanish by at least 4 months. In short, infant bilinguals were not confused.

Exposure to two languages early in life results in considerable plasticity in the system. One question that remains unanswered is how the tension between early specialization of the phonological system and later second language learning plays out. In the last chapter we discussed evidence that sensitivity to nonnative phonemes drops off in the first year of life. However, it appears that children can learn a second language without an accent. If this is the case, then there must be some plasticity remaining in the system.

To test whether young infants who are past the infant cutoff for recognition of nonnative phonemes could reacquire sensitivity to them, Kuhl and colleagues exposed 9-month-old infants to nonnative speech (Kuhl, Tsao, & Liu, 2003). However, they exposed infants to nonnative speech at the age of 9 months in both a real-life interaction and a passive audio or video listening format. In the live sessions, infants sat in front of Mandarin speakers for 12 sessions over a 4- to 5-week period. The speakers read books and played with toys using a natural tone of voice with the infants. The two control groups were also exposed to the language. However, rather than interact with a person they were either shown a video or listened to an audio program with Mandarin speech. By doing this, Kuhl could look at whether direct interaction with a person would make a difference in how quickly they were able to recover the lost speech sound contrasts.

The results revealed a strong effect for exposure to speech that occurs in social situations. Infants exposed to a live speaker showed sensitivity to Mandarin contrasts that were equivalent to those seen in infants who had grown up in a monolingual Mandarin environment. Furthermore, the effect lasted over a month. However, this effect did not appear in the groups of children who listened only to audio or just watched a video. Hence, direct exposure to a native speaker is important in order for infants to regain the sensitivity to nonnative phonological contrasts.

The interesting question that remains is what happens to bilingual children as they get older. Work by Laura Ann Petitto and colleagues (Petitto et al., 2012) tested a group of monolingual and bilingual babies ages 4–6 months and 10–12 months to a set of phonemes from their native language and a set of phonemes from their nonnative language. The patterns of neural activity were very similar in the younger age groups. Specifically, listening to both native and nonnative speech resulted in neural activity in the superior temporal gyrus, related to processing auditory speech, across both language groups at the young and older time points. Increases in the left inferior frontal cortex were observed in the older group but not in the younger groups. As expected, monolinguals showed a relative decrease in brain activity to nonnative speech sounds relative to native language speech sounds. Interestingly, sensitivity to both native and nonnative speech sounds remained in the bilingual group. This was especially evident in the brain activity seen in the left inferior frontal cortex, where bilingual older infants continued to show increased activity for nonnative speech sounds. Hence, not only does bilingualism afford the ability to distinguish two sets of speech sounds but it also appears to actually extend the length of time that a child can continue recognizing speech sounds from other languages. Petitto suggests that this forms a "perceptual wedge" that extends the length of time that a child remains sensitive to speech sounds.

So far we have reviewed evidence that suggests considerable plasticity in the infant phonological system. Not only can bilingual infants learn two phonological systems, but monolingual infants have the ability to regain lost sensitivity to nonnative contrasts. Given that infants can relearn a contrast, it seems very plausible that young children may be able to learn a second language in a very native-like manner.

PHONOLOGY IN A SECOND LANGUAGE

A number of studies have found that accent is related to the age of second language acquisition. Asher and Garcia (1969) conducted a seminal study in this respect. A group of Cuban adults who had come to the United States between the ages of 6 and 19 were recorded as they spoke in English. A set of raters then decided how native-like their accents where. The results revealed a strong relationship between degree of nativeness and age of second language learning. Since then a number of studies have been conducted that show a clear linear relationship between AoA and second language accent (Piske, MacKay, & Flege, 2001). Taken together these studies reveal that the later someone learns a second language the less they are going to sound like a native speaker.

If phonological development continues to occur across childhood and results in systems that are less and less able to adapt to a nonnative speech sound system, then we should expect that adults will also show these developmental effects. To investigate the link between language learning history and development of second language phonology, Susan Guion and colleagues have conducted a series of studies investigating the production of sounds. These results have revealed that the independence of vowel categories in bilingual adults is dependent on when they learn their second language (Guion, 2003). In an early study, a group of Quichua-Spanish bilinguals were asked to pronounce a series of vowels in both languages. Quichua is a native dialect spoken in the area near Quito, Ecuador. Spanish has five vowels: *a*, *e*, *i*, *o*, and *u*. Quichua only has three vowels, which would correspond to the Spanish *a*, *i*, and *u*. Linguists generally use a two-dimensional space to describe vowels. The first dimension corresponds to how open (i.e., English *a*) or closed (i.e., English *i* or *u*) a vowel is. The second corresponds to whether the vowel is relatively in front (i.e., English *i*) or in back (i.e., English *u*). Differences in the way that vowels are laid out in this space do appear across speakers of different languages. Figures of these vowels reveal a triangular shape, with each vowel in Quichua corresponding to each vertex. To accommodate two additional vowels, Spanish stretches the *i* and *u* further away making space for an *e* and an *o*. This results in a compressed vowel space in Quichua relative to Spanish.

Not surprisingly, monolinguals of each language will speak their nonnative language with an accent either because they are spreading the vowels out too much or because they are constricting them too much. Quichuan accents will appear in

Spanish because of the fact that the latter has more vowels. They will tend to produce a Spanish *e* and *i* or an *o* and *u* using a sound that lies between these two categories. Similarly, Spanish speakers will tend to slur Quichuan vowels by spreading them too far. In the case of Quichua and Spanish, accents appear either because of the fact that too many vowels are trying to be constrained into a space or too few vowels are trying to be expanded into the same space.

Guion tested a group of Quichuan-Spanish bilinguals who had learned Spanish at various ages. The results revealed a strong relationship between the AoA of each language and a particular participants ability to pronounce the vowels. Simultaneous bilinguals were able to pronounce vowels in both languages accurately. Bilinguals who learned a second language between the ages of 5 and 7 were close to native-like in both languages. As the age of second language increased, the ability to represent both sets of speech sounds diminished. The most extreme case of this was in the individuals who learned a second language between 15 and 25. In this group, there was a strong bias toward using the first language to guide pronunciation of vowels in the second language. That is, late learners tend to build their second language sound system around an already established first language system. Hence, the second language is more dependent on the first language in late learners.

Results from Guion's studies revealed that independence of the two systems is greatest in simultaneous bilinguals but much less so in sequential bilinguals. Specifically, those who learned a second language after 5 years of age exhibit differing amounts of dependence on the first language depending on the age at which the second language was learned. Childhood bilinguals show a mild dependence of the second language on the first language, whereas the late bilinguals' second language vowel categories are very dependent on the first language. In short, the older an individual is at the time they begin to learn a second language, the more their first language sound system is entrenched.

So far we have reviewed evidence that the ability to hear and produce speech sounds in a second language is related to the AoA of a speaker. Infant learners show incredible plasticity in the phonological system. They can learn to recognize sounds from another language even when that speech sound was lost during the first year. Children who learn a second language have softer accents than adult learners. One question that arises is how this dependence may play out in the recognition of phonemes by people with different second language AoA.

Despite this early entrenchment, there is evidence that later learners can recognize sounds in a second language. In one study, Flege and MacKay (2004) tested a group of Italians who immigrated to Canada between the ages of 2 and 13 (early learners) or between the ages of 15 and 26 (late learners). The results revealed that early learners in general could distinguish English vowel sounds better than late learners. However, there was considerable variability. For example, early learners who used

Italian often were found to differ from native speakers of English, whereas those who used it seldom were not. Furthermore, some of the late learners were able to distinguish these vowel sounds. Thus early acquisition does not appear to be a guarantee that native-like second language performance will be observed when recognizing speech sounds. Furthermore, it does appear that some late learners can learn to distinguish these sounds. The question that remains is how.

To uncover the mechanisms that differentiate early and late learning of a language, Pilar Archila, a graduate student working with me at the time, and Jason Zevin, a researcher with expertise in AoA effects and speech, asked precisely this question (Archila-Suerte, Zevin, Bunta, & Hernandez, 2012). Spanish speakers who learned English at different points in childhood were presented with two words and asked to rate whether they were similar or different. The sounds were chosen such that some English sounds were similar to sounds in Spanish (SAF and SEF) and others were much less similar (SOF and SUF). Pairs of sounds were presented, and people had to decide how similar the two sounds were using four different buttons that were marked as really similar, somewhat similar, somewhat different, or really different.

Archila's experiment was designed to uncover the role that AoA had in an individual's ability to detect differences across phonological boundaries. Specifically, she was interested in understanding if AoA played a role in determining how speech sound categories were formed.

One way to look at the data from each participant is to visualize it as having four different clusters, one for each sound. The amount of blur in each cluster indicates how well a person hears the similarities in the sounds. Imagine that the cluster is a single point. If this were true, then a person would hear every single variation of a sound as exactly the same. Reality is not nearly so clean. There is some variation in the way a speech sound is produced. Despite these subtle differences, native speakers recognize all these variants as the same sound.

Interestingly, the tightness of each cluster was correlated with second language AoA. Age of learning played a role when our participants had to decide that two items from the same category (SAF-SAF) were the same. Early learners, those who learned English before age 5, like monolinguals, show four tight clusters for each sound. However, intermediate learners, who acquired English between 6 and 9, showed distinct but blurry clusters; and late learners, who acquired English after the age of 10, showed even blurrier clusters. These results show that the age at which a person learns English plays a role in how well they can categorize two sounds as being from the same speech sound category. In other words, early learners, like monolinguals, are able to ignore slight variations in a sound in order to form a single sound category. Late learners use a very different strategy in which they use a relative comparison strategy to compare different sounds. Hence, as Flege and MacKay (2004) noted, late learners can be quite good at distinguishing between different phonological categories.

The behavioral findings from the study with second language learners was resonant with the sensorimotor hypothesis that Ping Li and I had proposed in the mid-2000s. Early development would make use of sensorimotor areas in the cortex to a greater extent. Late development was focused on ever expanding areas of the cortex. The fact that we found that AoA correlated with the ability to identify a phoneme (i.e., the tightness of the clusters) was consistent with this view. However, our evidence was indirect in that it did not involve direct observation of brain areas. What we needed was brain imaging data to more strongly make this case.

In Archila's next experiment, she set out to see if her behavioral findings actually translated into the brain activity predicted above. To do this she and Jason Zevin adapted the task used in the behavioral experiment by presenting participants a series of repetitions of the same two sounds (saf-saf, saf-saf, saf-saf, saf-saf) or two different sounds (saf-sof,...). As in her behavioral experiment, she tested monolinguals and bilinguals. For this task we used preattentive listening, in which participants are not asked to make any decisions about the stimuli. Furthermore, we asked participants to view a movie in order to make the time pass more easily and, perhaps more importantly, to keep them awake.

Under normal circumstances our fMRI study would have had fewer subjects than those we tested in the behavioral experiment. But these were not normal circumstances and we had both the foresight of Read Montague and the willingness of the University of Houston to pool its resources in order to pay a lump sum for scanning. Because of this, Archila was able to test 83 subjects total, 17 monolinguals and 66 bilinguals. The bilingual group's English second language AoA varied between 0 and 21 years of age, with their exposure to English varying between 24 and 3 years. The monolinguals had all been exposed only to English since birth and spoke no other languages fluently.

For years, I had wanted to look at the effects of AoA using the regression technique that we had first instituted on single words with Christian Fiebach and Angela Friederici in 2003. As you might recall, the logic behind this approach was to use statistical regression in order to calculate the effect of AoA taking out the effect of word frequency. Thanks to our large sample size we were able to take a similar approach, looking at this effect across individuals using second language proficiency and AoA as predictors of neural activity.

The results did not disappoint. Early learning of English was associated with increased activity when single speech sounds were repeated (same) or when the last sound was different. When the same sounds were repeated, early learning was associated with increased activity in the left superior temporal gyrus, which is very proximal to areas involved in speech perception. The effects of AoA were greater when the last sound was different. This was associated with increased activity in the left insula and a different portion of the left superior temporal gyrus. In addition, there was increased activity in a wide range of right hemisphere areas, including activity in the inferior frontal gyrus, the precentral gyrus, the rolandic operculum, and the superior temporal gyrus. Hence, listening to a change in speech sounds seems to activate a bilateral language circuit to a greater extent in early learners of a language. Listening to the same speech sound is associated with increased activity in brain areas that are involved in speech recognition to a greater extent for those who learned English early in life.

Given the strong effects we observed for early learners, we expected to see an equally large effect for late learners. This did not happen. Hence, learning a language later in life did not seem to lead to the recruitment of additional brain areas. You might wonder why this might be. Well, in fact, it turns out that learning a language later in life leads to increased brain activity, but that activity is driven by how well each person speaks the language. We'll discuss that further in a later chapter, when we consider the role that proficiency plays in the bilingual brain.

Taken together, the literature on the perception and production of speech sounds indicates a strong mostly linear relationship between age and phonological processing. There is no absolute cutoff. The loss of this ability declines slowly across age. Results from Pilar Archila's two studies are consistent with Patricia Kuhl's view on phonological development. Specifically, early learners form a set of native language magnets that serve to shape early phonological space. Archila's results show that early learners are much better at ignoring small variations in a sound to form a single phonological category. The second study shows that early learners' ability to hear the sameness of a speech sound is reliant to a greater extent on neurons in the superior temporal gyrus, which is associated with speech recognition. Contrasts in speech sounds also show this AoA effect. However, early learners go beyond the temporal lobe and recruit areas in the frontal lobe and many right hemisphere areas as well. In short, early second language learning relies to a greater extent on brain areas devoted to sensorimotor processing. Late learning comes to rely on a host of other brain regions, especially those in the frontal lobe in order to achieve the same goal.

GRAMMAR

In the last chapter we considered the link between grammar and phonology in the monolingual language acquisition literature. The same type of link also appears in the bilingual language acquisition literature. The early ability to learn two separate speech sound systems is followed by an equally early ability to learn two grammatical systems. Genesee (1989) and Meisel (2001) have found that bilinguals are able to produce sentences correctly even when the two grammars of each language conflict with each other. For example, Meisel has found that French-German bilinguals place their verbs in locations that correspond to each of their native languages just after their second birthday during the production of their first few small sentences.

Similarly, Paradis and Genesee (1996) have found that children place a negative after the verb in French (n'aime *pas*) but before the verb in English (do *not* like). In the case of simultaneous bilinguals, it appears that both phonology and grammar in each language are processed according to the language specific rules at a very early age.

The fact that separation of grammar and phonology occurs at such an early age in child bilinguals is consistent with the phonological bootstrapping hypothesis proposed by Demuth and colleagues. As you might recall, this hypothesis holds that aspects of grammar could be found in the underlying phonological and prosodic structure of a language. Evidence that bilingual children can differentiate between the two different sound systems from the beginning of infancy and later show differentiation of the two grammatical systems at the very beginning of learning fits in nicely with Demuth's hypothesis.

CRITICAL PERIODS IN SECOND LANGUAGE ACQUISITION

The early differentiation between grammatical and phonological processing also leads us to consider the nature of critical periods in grammatical processing. As we saw in chapter 1, the notion that age plays a role in the acquisition of a second language has its origins in Ribot's work in the late 19th century. As we have noted earlier, there is considerable evidence of early specialization of speech sounds, with a slow drop-off across childhood.

In 1989 Newport and Johnson published a paper that has become one of the most widely cited pieces of evidence in favor of critical periods in second language acquisition. In their study, a group of Korean-English bilinguals who had arrived in the United States between ages 3 and 39 (J. S. Johnson & Newport, 1989) were given a set of sentences and were asked to judge whether they were grammatically correct (Linebarger, Schwartz, & Saffran, 1983). Half of them had arrived before the age of 15 and half had arrived after the age of 17. The majority of these individuals were affiliated with the university either as graduate students, researchers, or faculty members. Hence, they were highly motivated to learn English and had a high level of education. Although Johnson and Newport used a variety of grammatical errors,[1] for our purposes I will discuss their general findings.

Results revealed a clear AoA effect. Early learners and monolinguals scored 269 correct out of a maximum of 276 possible. Those who arrived between ages 8 and 10 scored 256, and those who arrived between ages 11and 15 scored 236. Finally, those who learned between 17 and 39 scored 210. These scores drop pretty dramatically after age 7. In fact, the three later-learning groups scored significantly worse than the native and early group. The group that learned between the ages of 3–7 did not show any difference when compared to the native learners.

Based on the mean scores we could conclude that age 7 is the point at which it is no longer possible to achieve native-like ability in a second language. However, Johnson and Newport also wondered what the relationship between age and performance might be across the entire group. What they found was that age correlated quite highly with ultimate performance. However, this effect changed depending on the age range of the participants. Up until age 7, all participants scored as well as native speakers. Between the ages of 8 and 16 there was a strong correlation between age and performance on the test. However, between 17 and 39 there was no relationship between age and performance on the test. Johnson and Newport concluded that their results support a maturational account of language acquisition. Age matters during a time of great biological change. However, once a person enters into a mature state then ability plateaus and is no longer driven by the age of at which the second language was learned.

Johnson and Newport's study spawned a whole host of follow-ups in which people began to reanalyze the original data as well as add their own data. The story becomes a little more complicated once we start to take into account these other analyses. For example, we could ask, what happens if we cut the group in half at age 20 (Bialystok & Hakuta, 1994). Now both the late and early groups show a relationship between age of learning and grammar score. What if we look at a different group? Birdsong and Molis (2001) replicated Johnson and Newport's original study with Spanish native speakers. Their results showed the complete opposite effect. Early learners showed no effect with age but late learners did.

Taken together all these studies show a couple of points very clearly. First, age matters. Both studies with Korean and Spanish speakers learning English show that, in general, the older one learns the more likely that the grammar score will be lower. Second, the problem is that it is not clear WHEN age matters. Some analyses suggest that age matters for early learners but not for late learners. Other analyses suggest the opposite. These different analyses also suggest different critical periods, ages at which a second language learner can no longer reach native-like ability. By one measure it is 7, by another measure it is 16.

So far we have asked whether early learning has an effect on the ultimate outcome in language learning. However, there is another way in which early learning might have an impact. Specifically, we could ask whether adults exposed to a language early in life might retain certain aspects of that memory over others.

AOA AND ORPHANED LANGUAGES

As you might recall from chapter 1, I learned Portuguese at the age of 20, and mastered it to such a great extent that I could pass as a native. One could argue that this was due to the fact that I spoke English and Spanish and could use these two systems to "jerry rig" my way to the proper pronunciation. But there is, I think, another reason that I was able to learn these speech sounds so well. The reason is that Portuguese was in reality an orphan language.

During my childhood summer trips to Mexico, I would often listen to music. One of the most famous pop singers in Mexico was Roberto Carlos, a native from the northeastern part of Brazil. He had some success in Brazil but nothing like the huge following he had in Latin America, where his accent sounded exotic in Spanish sung songs. My dad used to like to go to the Spanish-speaking record store in the Mission District in San Francisco. I convinced him to buy Roberto Carlos records, to which I sang in Spanish.

On one of our record hunting excursions my dad found a record that looked just like the one I had at home, except that the cover was white not pink. Upon closer inspection it became clear that this record was the Portuguese version of the record I already had. My dad who would listen to the Portuguese hour on the radio decided to buy it for me. My curiosity piqued, I began to listen to these songs and soon enough I was singing them with a very thick Spanish accent. I probably sung to the record for about a year or two before I grew older and took on other musical interests.

That very thick accent remained for me when I took Portuguese as a college student and it did not go away during my first few months in Brazil. However, over time the thick accent disappeared entirely and I came to speak with the accent of a Paulista, as those from Sao Paulo, Brazil's economic capital are called. Gone was my thick Spanish accent.

Many years later I decided to sing a Brazilian lullaby from that Roberto Carlos album to my son Nikolas. And the day I sung it my regional accent stood in strong contrast to the Paulista that I had grown accustomed to as an adult. I realized that I sounded like a northeastern Brazilian, the same accent that Roberto Carlos had sung with in my childhood. All those years later, the early memory of that song had persisted and when I sang I sounded like Roberto Carlos. By that time I was an academic and I realized that it was as if Ribot had been reborn and the law of regression had appeared in a way I had never experienced before.

My experience was not unique. Anecdotally, the famous scientists Wilder Penfield (1965) found that having a German governess (i.e., live-in nanny) for his children during the first two years of life resulted in the apparent loss of an accent when his children reached adulthood. In laboratory tests, Tees and Werker (1984) found that exposure to Hindi sounds early in life led to better performance in adults in distinguishing sounds that exist in that language but not in English. Taken together these results suggest that early exposure to a language can form a permanent imprint that can be tapped later in life.

The results discussed above point to stronger resilience in the memories that accompany early life experiences. More recent work by Terry "Kit-Fong" Au and colleagues (R. Au, Joung, Nicholas, & Obler, 1995; T. K.-F. Au, Knightly, Jun, & Oh, 2002; Oh, Jun, Knightly, & Au, 2003) has sought to uncover how far these early childhood language memories may go. The results of Tees and Werker's study would suggest that early experience has an influence on the ability to detect sounds that are particular to a language. However, it remains to be seen to what extent early language exposure influences higher levels of language learning such as grammar.

In a first study, Oh et al. (2003) tested students of Korean who were in one of four groups: (1) Korean native speakers who immigrated to the United States after age 12; (2) individuals who had spoken Korean in early childhood but stopped by age 7; (3) those who had overheard the language but spoken it minimally; and (4) adult learners of Korean as a second language. Participants were asked to identify and say words that contained an English /t/ or a /d/ as well as words that had an intermediate Korean /t/ that is not present in English.

The results revealed that early exposure to a language does affect the ability to detect the intermediate /t/. All subjects who had been exposed to Korean in early childhood were better than nonnative speakers in detecting this sound. However, only the two groups that *spoke* Korean as children were able to *produce* it. A similar result was obtained for accent ratings. Those who spoke Korean in early childhood had a more native-like accent than those who did not. In short, hearing a language helps with hearing it later, but only speaking it helps with speaking it later.

But one question that remains is whether memory for speech sounds actually helps you to learn a language. In a separate study, Au and colleagues tested a group of Spanish overhearers who had been exposed to a language but did not speak it (Oh et al., 2003). Despite the fact that these participants did not speak the language, Au and colleagues found that overhearers had a much better accent than nonnative speakers. Hence, the advantage in speech without an accent appears even in those who do not use the language for a significant period of time.

The results looking at AoA with second language learning shows that age matters. Learning the phonology of a language is benefited by early exposure. The effect is so strong that children that are effectively cut off from a language can still recover that phonology later in life. As we will see a little bit later on, this early advantage for sound does not translate into the ability to process grammatical rules in the unused language. However, for speakers that get continuous exposure to a second language, processing of grammar seems to be more difficult for learners with later AoA. Interestingly, there is no definite cutoff age that marks a form of critical period for second language acquisition. The lack of a consistent cutoff suggests that the relationship between age and second language learning may not be as direct as those seen in the sensory literature. It may involve a number of factors. One of the most important is language proficiency, a topic we will cover in the next two chapters. I don't want to get too far ahead of myself, but there are a number of researchers who argue that the main driving force is language proficiency and not AoA (Abutalebi, 2008; Chee et al., 2001; Chee, Soon, Lee, & Pallier, 2004; Meschyan & Hernandez, 2006; Steinhauer, White, & Drury, 2009). But AoA has not disappeared entirely from the language map. AoA as a construct is still very evident when we turn to modern studies on the neural bases of bilingualism.

AOA AND THE BILINGUAL BRAIN

As you might recall from chapter 1, starting in the mid-1990s there was a wave of neuroimaging studies that reawakened researchers' interest in the bilingual brain. This wave was ushered in by two separate imaging modalities, positron emission tomography (PET) and functional magnetic resonance imaging (fMRI). Both of these methods are sensitive to the vascular response in the brain. Your brain is like any other organ in the body. It requires oxygen in order to function and is fed by arteries. When you engage in any task, the neurons in your brain begin to fire. For example, looking at an object in the world will lead to firing of the rods and cones in your retina, which are sensitive to grayscale and color light. These signals are transmitted through your optic nerve and back to your occipital lobe. The neurons in the center of the occipital lobe, known as the primary visual cortex, will begin to fire shortly thereafter. As the neurons fire, the vascular system will transport blood to this area. As the waves of neural activity travel across the brain, the cardiovascular system adapts and feeds the regions that have the largest and most sustained neural response.

The revolution of the 1990s allowed researchers for the first time to make connections between the vascular response in a particular brain area and a particular cognitive task. PET uses a "glucose bolus injection," a term researchers use to describe the process by which radioactive glucose is fed to the brain via an IV in a person's hand. As a brain area fires it takes up blood and glucose. Areas that fire more will have a stronger rate of decay. The radioactivity decay is picked up by a scanner. PET has fallen for the most part out of favor with most neuroimaging researchers because it is expensive, invasive, and cannot be repeated more than once a year. But it still represents the beginnings of the change that has transformed the field.

One of the first studies looking at bilinguals using PET was conducted by Daniela Perani and colleagues . In this study a group of late Italian-English bilinguals were asked to listen to stories in Italian, in English, or in an unknown language. When listening to stories in Italian, the native language, participants revealed larger areas of activity that extended from the temporal lobes up into Broca's area. Many of the same areas were active when participants listened to the second language, English. However, the extent of this activation was reduced. Perani et al. suggest that later-learned languages engage a smaller network than native languages when bilinguals listen to speech. Hence, the age at which an individual learns a particular language is reflected in the brain activity observed using PET.

The few studies using PET quickly gave way to those using fMRI. As I noted earlier, fMRI was a godsend for those of us interested in the brain bases of cognition. We could design an experiment, meet with an MRI physicist who would design a pulse sequence for the scanner, recruit subjects, take them to the scanner, and for a relatively modest sum come out with data that showed us how the brain reacted to our experiment. It did not involve the injection of radioactive glucose in order to track the flow of blood. fMRI was done noninvasively by tracking changes in the deoxyhemoglobin that accompanied the increase in blood flow.

The use of fMRI ushered in an increased interest in the brain bases of bilingualism. In 1997, when fMRI studies were just getting under way, Karl Kim, Joy Hirsch, and colleagues published a paper in the journal *Nature* (Kim, Relkin, Lee, & Hirsch, 1997). This fMRI study asked a group of early and late bilinguals to say "in their head" what they had done the day before. They were given a cue that would tell them to think about something that happened at a particular time of day (morning, afternoon or evening) and in a particular language (first or second). Researchers observed neural activity in both Broca's and Wernicke's area in order to look at whether each language showed two distinct areas of activity or just one glob.

In Wernicke's area both languages showed an overlapping area of activity in all the people tested whether they learned their second language early or late in life. Kim et al., however, found something different when comparing neural activity in Broca's area for early and late learners. In early bilinguals, brain activity for both languages was mostly overlapping. Brain activity for late bilinguals, on the other hand, showed two adjacent but clearly separate areas of activity. These results suggest that when reciting to themselves what they did the day before, early and late bilinguals invoke neural systems that are mostly overlapping in areas devoted to the comprehension of language. In areas devoted to language production, on the other hand, late bilinguals use separate areas for each language, whereas early bilinguals do not.

The findings from Kim et al.'s study carry implications for results that we discussed earlier. Specifically, the authors suggest that different areas of Broca's area may be a reflection of the specialization in producing language at an early age. Hence, early bilinguals may come to represent the motor programs for each language in overlapping areas because of the fact that they learn to speak both languages early in life. Late bilinguals learn one system first and then a second. Because a first language is already firmly established, the second language builds around it in adjacent but clearly distinct areas.

Despite the notoriety, the Kim et al. study leaves us with a couple of unanswered questions. The most important of these is what aspect of the task leads to differences in neural activity across these two groups. The task used by Kim et al., reciting to oneself activities performed the day before, could involve a number of aspects of cognitive processing that go beyond simple motor planning. This task involves a memory component in that each participant had to recall what they were doing the day before in a particular language. This task also involved self-talk in each language. Since we do not have a recording of exactly what each participant was saying, it is possible that participants may have produced a different amount or quality of speech in each language. Furthermore, it is unclear whether the retrieval of a memory in conjunction with having to turn it into a verbal code might have been responsible for the differences across groups. Finally, participants were producing sentences in each language. It is possible that differences in the way in which sentences were put together might have differed across groups and languages. Hence, we are not sure what aspect of language may have differed across languages.

Earlier in the chapter we considered two different aspects of language processing, speech sounds and grammar, that we thought might be influenced by the age at which a second language is learned. As we noted earlier, evidence that phonology is affected by the age of second language acquisition, can be seen in the behavioral literature. In 2010, work in our lab had begun to consider the neural bases of phonological processing in bilinguals. As we noted in the last chapter, if there is a connection between speech sounds and grammar then we should also find a relationship between AoA and grammar. To examine this possibility researchers have turned to looking at neural activity during tasks that require grammatical processing.

GRAMMATICAL PROCESSING AND AOA

The influence of AoA on grammatical processing has also been approached using neuorimaging methods. The earliest work was carried out by Weber-Fox and Helen Neville, then at the University of California, San Diego (Weber-Fox & Neville, 1996). For their study they recruited a group of second language speakers who had learned English between the ages of 2 and 16. They strapped the participants' heads up with an electrode cap and measured their brain waves while they viewed sentences presented one word at a time.

Work with electroencephalography (EEG) captures a very different form of neuronal processing. When neurons fire they send action potentials, which are responsible for the all or none activity seen inside the cell. Unlike fMRI, EEG relies on electrodes at the scalp that capture this electrical signal. A single neuron firing would not create enough of an electrical field. However, when thousands or millions of neurons are aligned in a similar orientation, the size of the field grows much stronger. These stronger fields can be picked up easily by the electrodes that are attached to the scalp. EEG has a number of applications both in clinical settings and research settings.

In order to uncover the timing of a particular cognitive event, researchers have used an averaging technique in order to arrive at a wave that allows them to see the neurophysiological signatures of a particular type of event. These event-related potentials have been used to make inferences about the timing of a number of cognitive processes. These different waves are seen as indices of different psychological processes. Of particular interest to language researchers is the N400, a negative wave that peaks at 400 msec, which is associated with deeper semantic processing (Kutas, Van Petten, & Besson, 1988; Lau, Phillips, & Poeppel, 2008).

ERPs have also been used to study grammatical processing. Unlike the processing of meaning, grammatical processing modulates brain activity at two different stages of processing. Researchers have found that grammatical errors generally result in changes in two brain waves at two distinct time periods. The first wave peaks at about 200 msec after a stimulus has been presented. It is a negative going wave that appears over left frontal sites and is hence termed the early left anterior negativity (Friederici, Hahne, & Mecklinger, 1996; Friederici & Meyer, 2004). This wave is commonly seen when listeners are presented with sentences that have phrase structure violations such as "The pizza was in the eaten." Because eaten is a verb and the context has built up such that we expect a noun (i.e., oven, box) the presence of a verb after "in the" leads to a phrase structure violation. Angela Friederici, an expert on the brain bases of grammatical processing, has proposed that this wave is part of the early automatic processing that goes into building a sentence structure (Friederici, 2002).

This first pass of grammatical processing is followed by an additional wave that is thought to involve more conscious grammatical processing. The second wave, a positivity that peaks around 600 msec, has been found to occur in sentences that have grammatical violations but also in sentences that temporarily lure you into the wrong interpretation (Osterhout & Holcomb, 1992; Osterhout, Holcomb, & Swinney, 1994). For example, the sentence "The broker persuaded to sell the stock was tall" leads people astray because a noun is expected after persuaded (i.e. persuaded me, the audience, or the seller). When people read the word "to" they need to reanalyze it in order to accommodate the fact that persuaded is actually a different tense of the verb (i.e., that was persuaded to). This particular garden path sentence occurs because there is a conflict between the two uses of persuaded, leading to a P600 wave at the word "to" because of the reanalysis needed. Taken together studies of grammatical processing in monolinguals show a very early EEG component associated with errors as well as a later component that serves to refine grammatical processing.

Weber-Fox and Neville sought to test whether these two components might differ in learners who differed in their age of second language acquisition. They asked individuals to read sentences that had grammatical violations such as "The scientist criticized Max's **of** proof the theorem." These sentences were designed to go wrong at the bolded word. Not surprisingly, learning English later in childhood modulated the size of early and late component EEG waves relative to monolinguals. The early negativity that appears in sentences with blatant errors in the parts of speech used (i.e., The scientist criticized Max's **of** proof the theorem) was reduced in all second language learners, even when the age of second language immersion occurred during the first three years of life. Furthermore, there was no indication of the brain signatures associated with reanalysis in individuals over the age of 11. For those who learned the language between the ages of 1 and 10, however, this later component was present. The presence of a later component suggests that individuals who learn a second language during early or middle childhood perform reanalysis in a similar manner to monolinguals. In short, the early components of the brain waves were reduced in all second language learners. The late components were reduced in those who learned a second language after the age of ten.

In 2011, Eric Pakulak and Helen Neville followed up this study by looking at the brain waves associated with grammatical errors in a group of late learners of English (who learned English between 10 and 12) who were matched to native English speakers in proficiency. Results revealed that native speakers showed both an early negativity and a late positivity for grammatical errors. Late learners, however, did not show such an early negativity. Furthermore, the late positivity showed a widespread pattern across the brain and extended for a longer time than that seen in native speakers. Similar results have also been observed by Manolo Carreiras, Horacio Barber, and colleagues at the Basque Center for Brain, Cognition and Language (Dowens, Vergara, Barber, & Carreiras, 2010). These two studies confirm that late learners use other mechanisms to process grammar relative to native speakers.

Taken together these three studies suggest that learning a second language, even during childhood, does not lead to a native-like pattern of electrical brain activity. The complete lack of an early negativity regardless of second language AoA suggests that automatic computation of grammatical processing may not occur in second language learners (for an alternative finding see Rossi, Gugler, Friederici, & Hahne, 2006). Hence, learning a second language at any age results in differences in the ability to compute early grammatical information. However, the later components of reanalysis necessary for controlled grammatical processing were present. Thus, detecting grammar errors in a second language relies to a greater extent on later reanalysis instead of early computation. Finally, there is some evidence that learning a second language later in life is associated with electrical activity in a larger brain network during reanalysis of grammatical errors relative to the network seen in native speakers.

In 2003, my colleague Isabell Wartenburger, along with a group of other researchers including Daniela Perani, wondered whether AoA might also influence the nature of brain activity measured using fMRI. Building on Perani's previous work (Perani et al., 1998), Wartenburger compared groups of Italian-German bilinguals that had learned their second language either late or early in life. Participants were shown sentences that had errors of case, number, or gender in German as well as errors in number or gender in Italian. The results revealed increased activity for late bilinguals relative to early bilinguals in the prefrontal cortex in a region that was just above Broca's area. This superior portion of Broadmann area 44 is typically associated with the need to retrieve the speech sounds of a word (Poldrack et al., 1999). Hence, it appears that late learners were having to access the speech sound of the article of a word in order to check whether it was correct. The early learners, on the other hand, showed no difference in the brain's blood metabolism when shown sentences with grammatical errors.

ISOLATING AOA

One issue that comes up with studies that have looked at the brain bases of bilingualism is that AoA is often confounded with language proficiency. In the case of the two studies just reviewed, Perani et al.'s study with language comprehension and Kim et al.'s study with production, there are potential confounds with language proficiency. These issues are part of a longer discussion taken up later in this book, so I will not belabor the point here. What is important for our purposes right now is to attempt to isolate the effects of AoA independent of proficiency. Traditionally, researchers have used two complimentary approaches in order to isolate effects of proficiency. The first approach involves the use of a statistical method in which the effect is minimized by using regression. The second approach is to create two groups that differ in their second language AoA but are identical in proficiency. This was the approach used by Wartenburger and colleagues.

Both of these comparisons have some limitations. When comparing monolinguals and bilinguals we are left with the possibility that any differences in the latter have to do with speaking two languages and not with having learned one language later. That is, it is unclear whether the difference between these groups might simply be due to the comparison between individuals who learned a second language and those who did not. In addition, comparing second language learners with different AoA does not resolve the question of whether a first and second language differ due to AoA. An important complementary approach would be to compare performance in a second language to that of a first language while matching for language proficiency across groups in both languages. Although Pakulak and Neville did this with ERPs in 2011, there was very little work exploring this approach using fMRI.

The opportunity to do this presented itself to my research team in 2002. With the help of colleagues in the Spanish department at the University of California, Santa Barbara, we were able to recruit late learners of Spanish with high proficiency in that language. When I realized that we could recruit a group of late Spanish learners who were dominant in English, a lightbulb turned on in my head. Work in my laboratory had noted a pattern of English dominance in a very different group of subjects. For many years, I had noted that Spanish native speakers who became bilingual in childhood would show up at our laboratory and score better in their second language, English. This dominance in English was found in tasks such as picture naming or word reading (A. E. Hernandez, Bates, & Avila, 1996; A. E. Hernandez & Kohnert, 1999; A. E. Hernandez & Reyes, 2002). Now I had the opportunity to compare native English speakers with native Spanish speakers in a way that was quite unique. Specifically, I could compare two groups that showed very similar language proficiency profiles but differed in AoA.

MAXIMIZING THE AOA EFFECT: GRAMMATICAL GENDER

The next task was to find a good language task that would help to differentiate the two groups of bilinguals. Up until this point most of my work had focused on the nature of word recognition and production. I had a few papers on the processing of grammar, and these mostly used a very high level task in which participants were asked to make decisions about who was doing what to whom in a sentence. As I transitioned from my PhD work into using neuroimaging, I began to search for tasks that involved single words. Ideally, I would choose a task that involved single words but was grammatical in nature. The task I chose came to me in collaborative work I had done just a few years earlier.

The nature of grammatical gender had been brought to my attention by Elizabeth Bates. Liz was one of the most verbally gifted people that I (and pretty much everyone who knew her) had met. I recall one incident in which Bob Buffington, the programmer answered a call from Liz and then turned to leave. He told me that he was going to "unfreeze" Liz's computer. Back in the days before webmail, sending e-mail involved opening a telnet window and logging into an account to send messages. Bob explained to me that about once a week Liz would freeze her computer because she typed faster than the terminal could update her keystrokes.

Liz's forays into foreign languages came with incredible fluency. She had learned Spanish in college and had given several talks in Spanish during lecture circuits. Later in her career she became a de facto Italian. She spent several months a year in Italy in an apartment she and her husband George Carnevale had bought in the Trastevere section of Rome. Liz mastered many things in the Italian language. But she struggled with grammatical gender. She stopped often times to ask her Italian colleagues the gender of a particular noun. She would then exclaim, "That makes no sense!"

For an English speaker the notion that grammatical items are masculine or feminine is grounded purely in semantics. Many nouns carry implicit semantic gender. For example, the word "boy" is not any different from "toy" but the former is more likely to be associated with males and the latter may (as in G.I. Joe) or may not (as in Barbie). However, semantic gender may also be marked explicitly on personal pronouns ("he," "she," and "it") or on nouns such as "waiter" and "waitress." The tendency to break up the world into masculine and feminine categories also extends beyond nouns that refer to sex-based gender. Interestingly, this bias also appears in children who have a tendency to judge artificial categories (i.e., buildings, elevators, and cars) as masculine and natural categories (flowers, trees, and birds) as feminine. The conceptual basis of nongrammatical gender also appears across languages (Boroditsky, Schmidt, & Phillips, 2003; Konishi, 1993).

However, certain languages extend the semantic nature of gender further. In romance languages, for example, the gender of a noun can be masculine or feminine and must agree with an adjective that qualifies it. This can be seen in Spanish phrases such as *el carro rojo*, which translates as "the$_{masc}$ car$_{masc}$ red$_{masc}$." People who speak languages that aren't gender marked, such as English, often wonder why grammatical gender even exists. Liz Bates argued that gender allowed a speaker to identify a noun without mentioning it. Spanish speakers may point to a pencil and a pen while exclaiming, *pasamela*, pass it to me. With enough proficiency, a Spanish speaker would pass the person a pen, *la pluma*, and not the pencil, *el lapiz*. In short, for a speaker of a gender marked language, Gender carries additional information. This information eases interpretation of sentences and serves to cut a potential word search in half.

The final dimension of interest was that grammatical gender, at least in Spanish, also has a regularity dimension to it. Nouns with regular gender carry an *-o* ending for masculine nouns (i.e., *carro*) and others carry an *-a* ending for feminine nouns (i.e., *casa*). Other words, however, that end in other letters (such as *s*, *t*, *z*, *n*, *r*, *e*) are much less regular. These words can be masculine or feminine. For example, *la fuente* (the fountain) is feminine, whereas *el puente* (the bridge) is masculine. Thus we could look at both regular and irregular grammatical gender to see whether the lack of direct connection between ending and gender mattered.

In a first study, we asked a group of monolingual Spanish speakers to judge whether a noun in Spanish was masculine or feminine while being scanned with fMRI (A. E. Hernandez et al., 2004). In every block of items, half of the items were masculine and half were feminine. In addition, we had blocks with regular gender signaled by an *-a* or *-o* ending and blocks with irregular gender markings that were signaled by words ending in *s*, *z*, *n*, *r*, *t*, *z*, or *e*.

Gender decisions regarding irregular nouns in Spanish led to increased neural activity in three distinct areas, the anterior insula as well as a superior and inferior portion of Broca's area, relative to regular nouns. In addition, there was increased activity in the anterior cingulate gyrus, an area that is involved in tasks that require increased cognitive effort. The role of the anterior insula in articulation has been well documented. Studies with patients that have difficulty with articulation, the physical movements associated with speech, have found that all of them have damage in anterior insula, which lies just inside the brain near Broca's area (Bates, Wilson, et al., 2003; Dronkers, 1996). Broca's area has also been associated with different aspects of speech. The lower portion has traditionally been associated with motor planning of speech (Broca, 1988; Graves, 1997). The upper portion has been found to be active in tasks that require people to access the speech sounds of a word. This area showed more activity when German native speakers were asked to produce the article (masculine, feminine, or neuter) when looking at the picture of a noun (Heim, Opitz, & Friederici, 2002).

How might we interpret these results? From a psycholinguistic point of view, subjects were engaged in deeper phonological retrieval and more complex motor planning while using more cognitive effort. Although this account sounds reasonable from a cognitive standpoint, there was something somewhat dissatisfying about it. It left me wondering what people were really doing and how that might help to explain the brain activity we observed in this condition.

In order to figure out what people might be doing members of my research team asked each participant to tell us about their experiences during the experiment. Most indicated that they found making gender decisions relatively easy. However, they did notice that some words were more difficult. For example, one subject noted that the when he saw the word *arroz* (rice) he searched for the gender marked article (*el*) and then made a decision. I then asked the subject what had happened for other words such as *carro* (car). Did he also produce the determiner in this case? This particular participant noted that he did not search for the determiner at all. Rather he would simply look at the last letter and then use it to guide his decision. Hence, decisions for each type of gender item follow two different retrieval paths. The path for regulars was fairly direct. Participants essentially had to look at a word's last letter and then make a decision. For irregulars, it involved producing the article in order to cue retrieval.

So let's revisit the interpretation of how regular and irregular items differ in their retrieval. Irregular items are more difficult because they require monolingual Spanish speakers to create a small noun phrase when deciding whether a word is masculine or feminine. This involves accessing the speech sounds, planning a motor response and preparing to articulate the word as revealed by the neural activity in the superior and inferior portions of Broca's area as well as the anterior insula. Finally, there was increased activity in the anterior cingulate gyrus that was indicative of the added cognitive effort required in this irregular condition. Hence, the brain signature associated with irregular items is substantially different from that observed for regular items. This is true even in speakers who have only been exposed to one language during their lives. Naturally, I wondered whether the effects of regularity might also tell us something about the bilingual brain.

COMPARING FIRST AND SECOND LANGUAGES

Grammatical gender does not come easily to English speakers. As I alluded to earlier, Au and colleagues wanted to know if the advantages for speech sounds also appeared for grammatical processing. To test this, they presented to a group of participants two objects ("a red cow" or "some red cows") and asked them to form a sentence with them. Au was making this task challenging because grammatical agreement in Spanish involves both number and gender. For example, "the red cows" would be *las vacas rojas.* Although the plural extends beyond the noun in Spanish, notice that it is marked by an *-s* similar to the way it is done in English. However, the overhearers also had to match the gender of the determiner, noun, and adjective. By looking at these two different ways to signal grammar, Au et al. could determine whether early exposure benefits grammatical functions that are not present in the dominant language. Interestingly, overhearers, like nonnative speakers, were unable to correctly identify gender errors (*la vaca rojo*) but were able to identify errors in number (*las vacas roja*). These results show that early overhearing may help with the basic elements of a sound system in a particular language. However, hearing a language does not help with all aspects of a language. In this case, grammatical rules that are particular to a language are not learned when a child is exposed to that language in a passive way during early childhood.

The work done in my lab looking at grammatical gender sought to exploit this difference between English and Spanish in order to understand AoA effects better. Since gender does not exist in its grammatical form in English, I could be sure that nonnative speakers would have a very different experience than that of native speakers. Specifically, native speakers would have early exposure to the grammatical rule, whereas native English speakers who learned Spanish in early adolescence would not. Hence, the fact that **grammatical** gender did not transfer easily from one language to the other allowed us to test the effects of AoA on grammar very directly (MacWhinney, 2004).

In addition to the native English speakers who learned Spanish in adolescence, we also recruited a group of early Spanish-English bilinguals. This group had early exposure to Spanish at home but extensive exposure to English at school. Over time education and immersion in the second language led to better language proficiency in English. This allowed us to match both early and late learners on English and Spanish proficiency. The greatest contributor to differences in brain activity would be the age at which Spanish was learned and not the particular proficiency in either language.

The results were simply stunning. When I first observed them I looked twice to make sure that I was not imagining them. Both groups showed increased activity for irregular items compared to regular items in the inferior portions of Broca's area. The stunning part was that both areas were adjacent to each other. The early Spanish learners showed activity in the area that was seen in monolingual Spanish speakers, one that is thought to be involved in motor planning. The late Spanish learners showed activity just below this. The activity was so extensive in the late learners that it spread into an adjacent portion that shows activity when people engage in retrieving the meaning of words.

These neural signatures clearly show that the time when a second language is learned has very different effects on people even when ability in that language is matched across groups. For early learners, processing of gender involves more extensive motor planning much like monolingual Spanish speakers. Late learners, on the other hand, need to engage a network that is involved in meaning retrieval. In some ways, this result sheds light on the comment made by Liz Bates about gender in Italian, which has a similar system to Spanish. It did not make sense. The reason it made no sense is that like the subjects in our studies and other English speakers, Liz was attempting to extend the semantic system to handle gender. The problem was that this semantic extension was a less efficient way to represent grammatical gender. In short, English speakers were taking the existing language representations that they already had and then adapting them to learn Spanish during adolescence. The late learners tested in this study did not differ behaviorally from our early learners.

So late learners can manage to retrieve gender correctly. However, their brains have to work harder and differently than that of native speakers.

So far we have considered the nature of AoA and its effects. What has become clear is that AoA has a definite effect on the processing of speech sounds and also on the processing of grammar. In terms of the brain, it is clear that early learning of a language is associated with early ERP waves and with brain areas that are more tightly tied to the processing or production of sound-based features. Thus as we saw in the monolingual and nonlinguistic literature, a second language learned early in life elicits a more sensorimotor form of processing than languages learned later in life.

NOTES

1. Some of the sentences had simple errors ("The farmer bought two pig at the market" or "The little boy is speak to a policeman") and others contained more subtle errors ("The man climbed the ladder up carefully").

4

PROFICIENCY, EFFICIENCY, AND EXPERTISE

IN THE LAST two chapters we considered the role that age of acquisition (AoA) plays in the processing of language. But as we noted at the outset, AoA is not the only player in the language-development game. A large body of research has asserted that development involves the process of becoming more familiar with something independently of when a particular skill was learned. This is also true in the aphasia literature that serves as a foundation for recent work on the bilingual brain. As you might recall, Pitres argued that it was the familiarity of a language, and not the age at which it was learned that played a strong role in how well it was spoken after recovery from neural insult. Pitres suggested that the strongest language would remain resistant to damage. This has come to be termed "Pitres's rule." In addition Pitres noted that many cases showed temporary "inertia" and that once that was overcome there was no need to think that aphasics were growing new brain centers for each language.

In the following two chapters, we will consider the issue of familiarity in the modern sense. However, modern researchers have used a number of terms to describe this in the last 100+ years. Despite the use of different terms, multiple literatures have explored the importance of familiarity in its various guises. In the domain of language, researchers use the term "proficiency" for bilinguals and to some extent for monolinguals.

Proficiency and familiarity, however, are not specific to the domain of language. Outside of language, researchers use the term "expertise" to describe how well a person performs a particular skill. These expertise effects have been explored across a vast number of domains. In fact, the nature of expertise has even been discussed in terms of the recognition of objects in the world. In extending Pitres's original observation, we will start with the general notion of expertise before proceeding to discuss proficiency as it relates to language.

EXPERTISE

The pioneer in studying the cognitive underpinnings of expertise was the product of an interesting melding of two seemingly unrelated interests, chess and psychology. Adriaan de Groot, the pioneer in this approach, was born on October 26, 1914, and grew up in Santpoort, a small Dutch village nestled between Amsterdam and the west coast of Holland. Because his family enjoyed chess very much, a great deal of his childhood was spent training to eventually become a world-class chess player. He competed for his country in the Math Olympiad on three different occasions for teams that finished in the top 10. In 1938 he finished fourth in the same tournament that had one of the best players in the world, Max Euwe.

In addition to their interest in chess, de Groot's family was also very well educated. This led him to spend a considerable amount of time in intellectual pursuits and gain considerable training in the math and sciences. Eventually, de Groot began to develop two different passions (Gobet, 2006). In addition to his interest in chess, de Groot also carried a strong interest in psychology. There was no undergraduate degree in psychology at the time in the Netherlands. So he chose to study physics and mathematics. During this period in his life he dedicated a lot of time to his studies while continuing to play chess.

As he became more engrossed in his graduate studies, he came to play chess less and less. But interestingly, chess remained on his mind, not as a sport but as a topic of inquiry. The first transition from player to observer of chess occurred when he served as a journalist in 1938 for a tournament in Amsterdam that featured the world's eight best players at the time. With the help of his friend Max Euwe, de Groot was able to test all of the great chess champions over several months. Little did he know that his discoveries in chess would lay the foundation for future work in expertise. For some people, it was the foundation of cognitive psychology as a field.

De Groot pursued this interest in chess in a dissertation on the subject. He asked players to describe their reasoning during the process of playing. In analyzing their descriptions, he uncovered a series of variables that could be used to describe how a person plays chess. This included the time to choose the best move, the number of different positions examined, and the number of different first moves considered. Interestingly, the differences in these variables were minimal even when comparing grand masters to novices. Hence, all chess players seemed to be considering the same types of information.

The fact that masters and novices seemed to be aware of the information was somewhat surprising. What was even more interesting was the fact that chess masters differed from less accomplished chess players in the way that they used the information available to them. Master chess players were able to hone in on the correct set of moves more quickly, could evaluate the success of a move better, and were able to organize these moves into a strategic plan. These qualitative differences between masters and novices were the bulk of de Groot's thesis.

The most enduring finding of the difference between expert and novice chess players appeared as a result of a short experiment de Groot performed. In this study, he asked players to look at a chess board for 2 to 15 seconds and to reconstruct it from memory. de Groot found that the number of pieces remembered on a chess board increased with chess playing ability. Max Euwe, the world champion at the time, could recall 22 pieces on a board, de Groot could recall 21, an expert player could recall 16, and an average player could recall 9. Hence, the ability to encode and maintain the pieces on a chess board in memory was related to chess ability.

de Groot found that memory for chess board locations also coincided with how quickly information was extracted. Even when shown a board very briefly, chess masters were able to extract information about potential threats. Most importantly, a chess master could quickly determine whether a particular position called for an attack or for defensive maneuvers. Hence, not only were chess masters able to recall many more pieces than novices, they could recall the gist of the position very quickly (de Groot, 1978).

Like many works that come before their time, de Groot's dissertation lay dormant for many years. During this time, he conducted a series of studies on psychometrics and testing and published an important book on methodology. It was 20 years before this topic would come to the forefront of cognitive psychology when his student Riekent Jongman chose it as his dissertation topic. In his thesis Jongman found that eye fixations for masters and novices differed substantially. Specifically, chess masters' eye movements were smoother and faster and could zero in on the particular position of a piece very quickly. The final part of Jongman's thesis focused on the way in which redundancy was treated by master chess players. Specifically, Jongman found that chess masters could determine the position of a piece by figuring out its relationship to other pieces. Since each piece's position depended on the other pieces, there was no need to encode each piece separately.

The fact that expert chess players could use redundancy suggests that the benefits might be specific to situations that experts encounter often. To investigate this hypothesis, Chase and Simon (1973a, 1973b) asked chess experts and novices to study chessboards and remember the layouts, which might or might not be encountered in actual play. The results revealed that expert performance was restricted to layouts of real chess situations. If the chess pieces were placed in patterns not used in real chess games, memory in chess experts and nonexperts did not differ. The fact that chess masters' expertise is relevant for real chess layouts suggests that their expertise is restricted to what they have experienced. That is, chess experts are not remembering more pieces. Rather, chess masters encode patterns and configurations. It is almost like when a native speaker of any language can recognize letters better when they are in words than when seeing them alone (Reicher, 1969; Wheeler, 1970). These "chunking" effects are present across many domains and usually arise from extended practice (Gobet et al., 2001).

The pioneering work of de Groot and the follow-ups by his students and then later by Chase and Simon offer us an extension of Pitres's rule. They do this by further refining the notion of familiarity. In the case of chess, the more familiar someone becomes with a chess board, the more likely he or she is able to see each configuration as a pattern.

Chess is not something that people learn from birth. It is clearly an acquired skill that is learned well after the first birthday. Yet, experts develop types of intuitions about chess that could be applied to other patterns. In fact, chess players' exceptional ability is really an extension of pattern perception, a domain that is part of every human's cognitive toolkit. To find perceptual expertise, we just have to find a domain that involves lots of practice. Luckily for us, such a domain exists. Just look in the mirror and you will encounter the most common pattern of our everyday lives.

PERCEPTUAL EXPERTISE

Humans are facial experts. The average person is born with enough visual acuity to distinguish the facial features of the person cradling him or her. This facial ability continues across our lifetime. And for good reason. Faces carry a lot of important information. They tell us about a person's identity, gender, and emotional state. Faces are so important that the inability to recognize them can have a devastating effect. Of particular interest is the fact that the ability to recognize faces can be lost in individuals who suffer from brain damage. The loss in the ability to recognize faces had been documented in the 19th century by Hughlings Jackson, who described the case of a man who could not recognize his wife and who could not navigate his way back home (Hughlings Jackson, 1876). From these deficits, he argued that the loss was not in the ability to recognize objects. These patients could see objects. The problem was in their ability to match the visual form of an object to the form in their memory. Hence, they were unable to put them into the appropriate visual category.

Despite the pioneering work of Hughlings Jackson, the most famous case study at the time was from Charcot, who presented the case of Monsieur X., who had been born in Vienna and spoke several languages well aside from his native German. He had also enjoyed and used visual imagery to a great extent. For example, imagery had been a method that helped him to remember many of the passages of the Iliad. So when his visual imagery was impaired, M. X. noticed it very quickly. He reported that images around him appeared new and strange and that his visual memories of shapes and colors had disappeared completely. He had also lost the ability to recognize his wife and his children. The visual world was so unfamiliar that he had even been seen apologizing to himself in the mirror. His ability to recognize faces was gone (Young & van de Wal, 1996).

Building on the cases that Hughlings Jackson, Charcot, and others had reported, Otto Pötzl, one of the first to suggest that bilinguals might have a language switch, had also produced an important review on visual agnosia (Lange, Pötzl, Brown, & Liepmann, 1988). "Agnosia" was the name given to the inability to recognize objects due to brain damage. Whereas work in the 19th century had brought these cases to light, Pötzl provided a classification system that included four different types of agnosia. These four types included difficulty seeing objects either real or in a picture, difficulty seeing events, difficulty reading words, and difficulty with orientation. This classification system ushered in a third wave in the study of agnosia, which sought to provide a more accurate description of the syndrome.

It was during this third wave of agnosia research in the mid-20th century that Joachim Bodamer, a German neurologist, made a very important new addition to this emerging classification system (H. D. Ellis & Florence, 1990). Interestingly, this was not due to the experiences he had in the clinic. Rather, the contribution came because of his work as a doctor during World War II, when he treated troops in Russia, France, and Bulgaria, many of whom had suffered brain damage due to gun shot wounds. It was during this time that Bodamer encountered S, a patient who would usher in a new subclassification in visual agnosia.

On November 9, 1944, S was admitted into the hospital for treatment. At first completely blind, he slowly recovered his vision across a number of different phases. During the first phase S could not detect color. The world was purely black and white. Interestingly, his loss of color vision was purely restricted to real life. S reported being able to dream in color, and gradually his ability to perceive color during his waking hours came back. Second, he transitioned to a phase in which he had difficulty seeing the contours of objects. This led to a difficulty with perceiving objects, especially in scenes with multiple things. He also reported having difficulty reading.

Bodamer also noticed that S had a profound problem with recognizing faces. He could recognize all the components of a face, but the faces had no meaning to him. This difficulty even appeared when trying to recognize faces that he saw all the time. Self-recognition was very difficult. When staring at a mirror, it would take him an inordinately long time to recognize his own face. This task was impossible when he had to choose his own face in a group photo. He could distinguish men and women by the way in which their hair was arranged. But even using these distinct characteristics he was not always accurate. Finally, S had difficulty with animal faces. He might recognize the pointy ears of a rabbit, but S stated that the faces of animals never materialized. S seemed to be relegated to recognizing faces piece by piece. The problem is, as we'll see later, face perception is a wholistic process. Doing it piece by piece is inefficient.

To account for their deficit, Bodamer used the Greek word for face, *Prosopon*, to come up with *Prosopagnosie* in German. The coining of the term and his initial observations have served as the starting point for considerable research since then (Farah, 2004). Despite the fact that face perception seems to be the greatest impairment in prosopagnosia, there does appear to be residual impairment of other visual abilities. For example, patient S had problems with seeing colors and with reading. Hence, prosopagnosia tends to go together with other deficits (Farah, 2004).

In the cognitive psychology field face processing has also received attention, which builds on the work in the field of neuropsychology. One of the most intriguing findings is that face processing changes when faces are placed upside down. These inversion effects lead to a breakdown of the wholistic processing of faces. The most dramatic demonstration of this is the Thatcher effect named after the former prime minister of the United Kingdom, who served from the late 1970s up until the end of the cold war in 1990 (Thompson, 1980).

Thatcher was known as a tough politician. If you look carefully at the figure you will notice that each picture is slightly different. But for the most part people don't really consider these differences to be major. For a very different picture (pun intended), turn the book upside down. What you notice is a ghastly difference

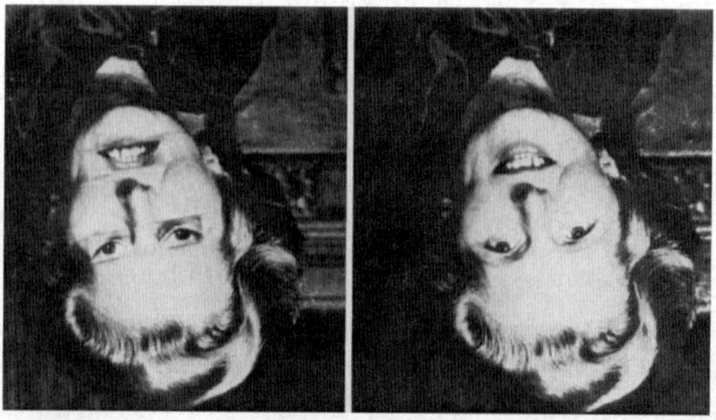

FIGURE 4.1 The Thatcher effect named after the former prime minister of the United Kingdom. Turn upside down for full effect. First published in Thompson, P. (1980). "Margaret Thatcher: A New Illusion" *Perception, 9*, 483–484, Pion Ltd., London, www.envplan.com.

between the two renderings. Perhaps you could say that one is the image seen by her supporters and the others by those who viewed her less favorably.

Work in the neuroimaging literature has further defined the basic brain areas that modulate the processing of faces. Of particular importance, is small section of brain that lies at the juncture of the parietal and occipital lobes called the fusiform gyrus. In the early 1990s, Justine Sargent and colleagues (Sergent, Ohta, & Macdonald, 1992) asked participants to look at a set of black-and-white images. This included faces or sine wave gradients (black and white bars) under a passive viewing condition. In addition, participants were asked to decide if faces were male or female, to determine the identity of a face, or to decide what object was being presented to them. The results revealed that the fusiform gyrus was critical for the processing of faces. This has also been found in studies using fMRI (Kanwisher, McDermott, & Chun, 1997). These studies, at least on the surface, support the view that face processing is a highly specialized process which requires specific brain areas.

The finding that face processing is highly specialized and depends on very localized brain areas leaves us wondering exactly what the underlying processes might be. Specifically, we are left with the question of what exactly makes faces special. One possibility is that faces have such importance that they have their own specialized system, which has been perfected over centuries of evolution. But there is a second alternative, one that fits in nicely with the importance of expertise. Isabell Gauthier and colleagues suggest that face processing is just another form of extreme expertise but not for the reason most people might think.

Earlier we discussed the Thatcher effect, which has been taken to indicate a specialization of face processing. Viewing of faces in a right side up manner leads to configurational viewing, hence the ghoulish face. However, turning a face upside down leads to part-by-part viewing and the face does not look ghoulish at all. The problem is, as Gauthier points out, that many other types of stimuli also show a Thatcher effect. If you look at the words "Margaret Thatcher" upside down, you may notice some subtle differences between the two versions. But now take the book and turn it over again. Once again you notice that the two versions look different when they are seen right side up.

Gauthier, as a strong advocate of an alternative view, has proposed that the behavioral signatures of face processing arise because of the extensive number of hours that people spend viewing faces (Wong, Twedt, Sheinberg, & Gauthier, 2010). Using this logic, Gauthier has set out to look at these inversion effects in domains where there

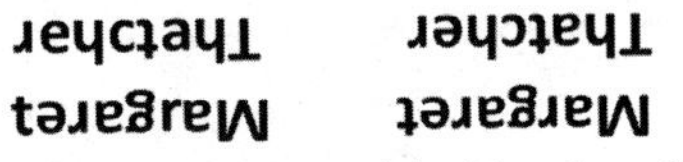

FIGURE 4.2. A less dramatic but effective version of the Thatcher effect.

is extensive expertise. For example, bird and car experts will show the same inversion effects that are seen for faces in nonexperts. These experts also show increased activity in the fusiform gyrus in the same brain area that shows increased activity for faces (Gauthier, Skudlarski, Gore, & Anderson, 2000). Furthermore, individuals who are trained to visually recognize greebles, small creatures that have complex body-like configurations, start to show inversion effects after several days of training (Gauthier & Tarr, 2002). Neural activity when viewing these figures also show increased effects in the fusiform gyrus as expertise improves.

In summary, the buildup that leads to the behavioral and neural signatures of face processing are also present in other visual categories in which expertise is gained. One way to think about this is to think of humans as face experts. Think about it, children see dozens of different people a day, and may encounter many of them multiple times. If we had a face counter attached to each child, it would very quickly reach a million faces. Imagine some other visual category that we have encountered as frequently. In short, our face expertise arises from a prolonged and intense form of practice. Hence, damage to relatively small areas of the brain can lead to prosopagnosia. Perhaps you have never thought of recognizing faces as being due to practice. If so, then it is about time you met someone who thinks everything has to do with practice. His name is K. Anders Ericsson, and he thinks that all forms of expertise have to do with practice.

WHAT MAKES EXPERTISE?

K. Anders Ericsson was born in Stockholm in 1948. He grew up in a middle-class family that valued hard work and effort. He recalls his parents telling him that anyone could achieve whatever they wanted as long as they put in the requisite time and effort. The value of effort and time also played a role in his approach to studying. For example, in high school, his classmates would often memorize important dates in Swedish history. Ericsson took a different approach. He would read a few books on the topic so that he could learn more about this particular period in history. This allowed him to reconstruct the dates in his head (Gare, 2009). So rather than having to memorize a few disjointed facts, he had learned a much longer narrative that gave his memory context. By using this larger structure, Ericsson could reconstruct all the answers to a test accurately and efficiently.

After high school, Ericsson enrolled at the Royal Institute of Technology in Stockholm. As he became more and more interested in thinking and problem solving, he turned to psychology to pursue his interest. He was very interested in using verbal reports, a form of data collection that was not highly valued by psychology faculty at the time. However, due to his strong scientific background he was allowed to use this methodology anyway. In his doctoral dissertation, he uncovered how verbal reports were able to capture problem-solving processes. In 1976 he received his PhD in psychology from University of Stockholm and began to look for a place to continue his training in psychology.

He followed his dissertation work by taking up a postdoctoral fellowship at Carnegie-Mellon University, where he worked with Herbert Simon and William Chase. Herbert Simon, famous in his own right for his work that bridged economics, psychology, and computer science, thought of the expertise literature as an extension of his interests. With him Ericsson developed even better measures to accurately capture the ways in which experts reason about problem solving (Ericsson & Simon, 1980, 1993). Chase was the one who had really focused on the issue of expertise and extended it to looking at the nature of chess players' ability. Ericsson came to this environment with a profound interest in how expertise developed.

His stay at Carnegie-Mellon yielded two beneficial outcomes. First, it allowed him to establish clear protocols for analyzing the thinking processes underlying problem solving (Ericsson & Simon, 1980, 1993). Second, this gave him the methodological tools that would later yield interesting data. With Bill Chase he extended his experimental techniques by examining whether certain cognitive measures of people's ability could be altered by practice. To do this he sought to extend people's ability to conduct the digit-span task.

How many things can you keep in memory for a short period of time? If you are like most of my psychology undergrads, you would probably gravitate to an answer of 7 ± 2. This is based on the digit-span task in which people are asked to keep a number of digits in memory for a short period of time. Ericsson thought that with extensive practice, the working memory span of most people might be expanded significantly past 9. In short, his view is that memory for digits is subject to change with training.

To illustrate why Ericsson might think this is true let's take a simple example. Imagine I give you the following letter string to memorize: jkljrngfjcrrgbbcgbbo. For a nice demo, copy it onto a piece of paper, close your book and try and memorize this string. If you do so, you'll notice that it takes several minutes if not longer to memorize this quickly. But you can remember this string rather easily with a couple of simple assumptions.

Let me give you a hint, try thinking about the two initials of every American president from 1960 to 2016. I left out the F from JK to make it more opaque. But if you go through this exercise you'll notice that you can almost recite the entire string from memory. You could probably get very fast very quickly as you tie in each of the letters with the memory of a president. What you have done is leverage your long-term memory as an aid to help extend your short-term memory. The important thing to notice is that the ability to store information for a short time is dependent on your underlying knowledge base.

For Ericsson, expertise in a particular field such as chess, music, mathematics, or poetry uses the same underlying cognitive processes that allow all of us to quickly memorize jkljrngfjcrrgbbcgbbo. He argues that just as you and I can leverage general knowledge of the world by using long-term memory, experts learn to leverage their memories in a similar way. As experts encounter more and more instances in their domain of expertise, they will begin to create a domain-specific vocabulary that consists of larger and larger chunks. Chess players no longer see single pieces but rather see configurations much as we see faces. Writers will begin to see entire books in their heads. Architects will close their eyes and imagine an entire blueprint. The question that arises is how. How is it that experts attain their abilities?

One possibility is that extraordinary individuals are born with talents that no one else possesses. Historically, there are many stories about extraordinary achievements. For example, Mozart is often credited with having composed his first symphony at the age of six. There is also the story of Paganini, a 19th-century violinist who was so good that he could continue playing even as one string after another broke until he was left playing a single-stringed violin. Ericsson discounts these stories as bordering on mythical. Mozart had a lot of help from his father and most likely really produced his first symphony at age 21. He can find no documented evidence of Paganini's one-string incident. In short, these individuals were often credited with extraordinary achievements that were not entirely accurate.

Further evidence for the view that expertise does not result from extraordinary circumstances or result in extraordinary abilities comes from measurements taken outside of the particular domain of expertise. Ericsson has found no evidence that experts are globally different from nonexperts. First, experts and nonexperts do not differ in IQ or other measures of general intelligence (Doll & Mayr, 1987; Taylor & Getzels, 1975). For example, at a press conference in 2006, after the year-end championships, Roger Federer, the holder of the most major tournament wins in tennis, was asked about how he became so good at tennis. He talked about how he spent the early part of his career getting a feel for the game. He would play many different types of players and won and lost many times. This information helped him to improve tremendously. The very next question posed to him was whether he got help from himself or Dr. Freud. Federer responded "Don't know him. Who is he?" Here is a man who has inspired some to view tennis as a religious experience (Wallace, 2006). However, this incredible artistry on the court does not translate into knowledge of the history of psychoanalysis.

A second trait of experts is the amount of time that they spend on their tasks. Ericsson estimates that expertise arises after approximately 10,000 hours of practice

(Ericsson, Krampe, & Tesch-Römer, 1993). However, not all hours are equal. There are many people who spend multiple hours at their task but do not seem to improve. For example, Camerer and Johnson (Camerer, Johnson, Ericsson, & Smith, 1991) found that very experienced diagnosticians such as radiologists can perform worse than professional and graduate students. Similar findings have been reported in other professions that involve assessment, such as clinical psychology (Dawes, 1996). Hence, absolute time on a task is not the only predictor of success.

Some professionals, however, do show improvement over time in a number of domains. For example, surgeons continue to improve as they get older. What is the difference between a radiologist and a surgeon? The most important factor, according to Ericsson, is that a person engage in deliberate practice. This type of practice has clear goals and involves feedback on performance. Surgeons know when things go wrong and must adjust when necessary. There is constant feedback from the environment in terms of how to perform the task. However, diagnosticians often times never find out if the tumor they saw was indeed cancer or whether a treatment really worked. Ericsson argues that feedback is essential for people to continue showing improvement. The most crucial aspect is that practice involve a goal and that feedback be given about whether the goal is reached or not. In fact, the best radiologist Ericsson ever encountered would always follow up his diagnoses in order to verify their accuracy. He continued to improve for many years afterward. Through deliberate practice, this diagnostician was able to overcome the limitations that held many of his colleagues back.

DELIBERATE PRACTICE AND LANGUAGE

If in fact the concept of deliberate practice applies across many different domains, then there should be evidence that differences in language proficiency can be attributed to differences across groups. One small caveat to consider is whether language involves training. Specifically, it runs counter to strong nativist views that suggest that biology is the prime force in language development.

One principal factor that has been found to play an important role in the development of language is the amount of speech that is directed at the child. Variation in the amount of speech that a child hears can occur for a number of reasons. It is known that children from different cultures receive different amounts and types of linguistic input. For example, children from families in middle-class industrialized nations are often spoken to directly. To accomplish this, adults often modify their speech when addressing the young infant. The modifications that occur in infant-directed speech help these children hear single words more easily. By biasing the intelligibility of single words, adults guide infants into quickly uncovering word boundaries, which leads to early production of words and eventually to the production of word combinations. Children who are not spoken to directly, on the other hand, tend to memorize chunks of speech. Since these chunks are not based on building a word-based vocabulary, they are much less fragmented. The learning of larger memorized chunks of speech by children who are not spoken to directly results in language that is much less analytical.

Evidence in favor of later language development has been reported in cultures where adults do not directly speak to children. For example, the speech of rural African American children (Heath, 1983) and in child second language learners under extreme immersion conditions (Bates, Bretherton, & Snyder, 1988; Wong Fillmore, 1991) tends to be less analytical than that of those from industrialized middle-class families. Similarly, in groups like the Walpiri of Australia and the Mayan children of Mexico children are also not spoken to directly. Children in these cultures appear to show later language development relative to children from North American middle-class families (Bavin, 1992; Brown, 2001). Hence, culture influences the type of speech that children hear and ultimately plays a role in the nature of language development.

In addition to differences across cultures, there are also differences within cultures that play a role in the language environment faced by children. One factor that has been found to play a role across all cultures is socioeconomic status (SES). The amount of speech that children hear can vary substantially across families that vary in SES. Single mothers who live in public housing are often described as living in households that are practically devoid of maternal speech (Heath, 1990). In another study, Hart and Risley (1995) collected samples of conversations inside the homes of low-SES (on public assistance), mid-SES (working class), and high-SES (professionals) families that had a child under the age of 2. The results from these analyses revealed vast differences in the number of words heard by children in high (215,000 words), mid- (125,000 words), and low-SES families (62,000 words). Furthermore, the higher the SES of a household, the more likely that a child would hear a larger variety of words. In short, SES leads to a greater exposure to words and to the types of words that a child is likely to hear.

In addition to the quantity of speech, there are also qualitative differences in the type of speech that children from homes with different SES levels hear (Hoff, Laursen, & Tardif, 2002). The speech of higher SES mothers was more frequently uttered for the purpose of eliciting conversation than the speech of lower SES mothers. Finally, a lower SES mother's speech was more likely to be directive of her children's behavior than the speech of higher SES mothers. Hence, the quality and quantity of speech varies across different levels of SES.

SES as a potential explanatory factor was also considered by Pakulak and colleagues (Pakulak & Neville, 2010). Specifically, they looked at the extent to which SES played a role in neurophysiological underpinnings of language processing. On its own, SES had only a moderate relationship with EEG measures in the anterior sites for the late anterior negativity. However, this relationship became stronger when controlling for the number of hours that participants read, were read to, and the amount of time they spent playing video games. SES does exert an effect on the brain waves associated with syntactic processing in a sentence context. Furthermore, it appears that this effect is independent of the number of hours that were spent performing various language related activities. In conclusion, it appears that it is not just the quantity of time that is spent using language that plays a role in differences in adults. Rather, it is quality of this input. More specifically, children from higher SES families receive considerably more feedback in their conversations with adults. Although adults may not be explicitly instructing them about language, by biasing specific types of interactions high SES parents are continuously guiding their children with richer and more diverse language interactions.

The effects of this interaction can be seen in studies that have aimed to "close the gap" between children of high- and low-SES families. In one study, Laura Justice and colleagues exposed a group of preschool children to a language enrichment program (Justice, Mashburn, Pence, & Wiggins, 2008). Considerable gains were seen in children from low-SES families. However, this benefit was lost when children stopped attending the program. Children from low-SES homes can show improvement in programs that mimic the environment seen in high-SES homes. However, like most of the expertise literature, it takes quite some time before the magical 10,000 hours can be achieved during a regular school day. When children's attendance becomes sporadic, the effects of enriched exposure to language begin to disappear.

THE CHARACTERISTICS OF EXPERTISE

So far we have discussed how experts acquire their abilities through deliberate practice, which involves feedback about performance. Furthermore, we have noted that these expertise effects are specific to their domain of mastery. Experts in general do not have higher IQs than the general population. This helps to elucidate the underlying process of gaining expertise.

Anders Ericsson also alludes to the process by which experts gain their advantage. Recall, that in training people to remember long strings of digits, he suggested that individuals began to create traces in long-term memory that they could retrieve as wholes. Thus experts come to represent knowledge differently than novices. An important question is how this knowledge differs from that of novices.

In chapter 2 I discussed work by Sian Beilock on expertise. As you might recall, expert and novice golf players were asked to either putt a ball under conditions that required them to explicitly keep track of their body movements or they were asked to perform the putt while listening to a set of words they would later have to recall. Results revealed that performance in experts dropped when monitoring their movements but not when monitoring the words they were hearing. Novices, on the other hand, showed the exact opposite pattern. They were better when monitoring their performance but worse when asked to perform a secondary task.

These results point to an interesting dissociation in the way that experts and novices process information. Novices tend to use explicit memory when accessing the various components of their putt swing. Because they are in an earlier stage of learning, novices have to think consciously about what they are doing. Experts, on the other hand, have spent a considerable amount of time practicing thousands of putts. This practice leads to the putt becoming an automatic whole, a motor chunk. When experts become really good at something, they are able to seemingly do multiple things at once. The ability to pull up an entire motion as one piece comes with a small price. Peeling it apart consciously actually hurts their performance.

To illustrate how these motor chunks are difficult to peal apart, I often ask my cognitive psychology class to try a simple thought experiment. I wanted them to tell me how to walk. So I say, "Imagine I was an alien from another planet and I wanted to learn how to walk. What would you tell me?" They usually tell me to put one foot in front of another. So I do that and begin to stomp across the floor. They agree that no one walks that way but that it would work. I keep prodding them to give me a better way to walk. Finally, I stop and suggest that they should step from heel to toe. They all realize immediately that they had left this important part of the step out of their descriptions. As expert walkers they were having a very difficult time breaking up the steps (pun intended) that were necessary for walking.

A dissociation between the two forms of memory termed "declarative" (what we can talk about) and "procedural" (what we can do) was at the root of my students' inability to access the relevant information. Like expert golf putters, the expert walkers in my class had a very difficult time telling me how to do something they do automatically. In fact, it almost seems as though we are completely unaware of what our body is doing while we walk. The difficulty that people have in accessing information about highly practiced skills such as walking, suggests that these types of skills may involve different neural systems than those we can talk about. Work in the neuropsychology and neuroscience of memory has uncovered one of the most striking dissociations between declarative memory and procedural memory. The most fascinating example of this dissociation is the patient HM, a classic example of a person who showed a stunning dissociation between procedural and declarative memory.

LEARNING WITHOUT KNOWING

Henry Gustav Molaison, known as HM to protect his privacy until his death, was born in 1926 in Manchester, Connecticut, to an Irish mother and a father from Louisiana. He grew up an only child with no adverse events to report until the age of 9, when he was knocked over by another bicycle rider and hit his head. Within one year of his accident he began to experience seizures that increased in intensity and length. At age 16, he suffered his first grand mal seizure. His seizures continued to increase in frequency even when treated with antiseizure medication.

At age 27, unable to hold down a job or lead a normal life, HM learned of a new procedure that might alleviate his intractable epilepsy (BBC, 2010). Previously, we discussed the procedure by which William Penfield had started to treat epilepsy. By 1953 news of this procedure had spread in the neurosurgery community. Dr. William Scoville reasoned that this surgery might help HM. On September 1, 1953, Scoville removed HM's hippocampus as well as other surrounding areas. The hippocampus is a small structure in the interior portion of the middle temporal lobe. If you point your finger right next to the side of your eye socket and then slide it to just in front of your ear lobe you will point in the vicinity of the hippocampus. It lies inside the brain, not at the surface.

The removal of HM's hippocampus treated the disease successfully. His epilepsy was gone. He seemed for the most part to interact normally with those around him. His memory also seemed intact. He could remember things about his past quite well and could also remember a phone number long enough to dial it.

When Penfield and Scoville were performing their pioneering work, Brenda Milner a neuropsychologist from the Montreal Neurological Institute became interested in the potential deficits that might arise from this operation. Milner began traveling to Hartford to study HM before and after the surgery. Together with Scoville, Milner had spent a lot of time informing HM of the surgery. Across multiple trips, HM had learned that there might be some neurological damage and entered into numerous discussions with his parents. Most importantly, HM got to know both Milner and Scoville before the surgery.

Shortly after his operation, Milner traveled to see HM. He remembered her quite well and acknowledged the operation and the many discussions they had had. Despite his intact memory, Milner noticed that HM was not the same man after the surgery. He had changed. Even HM himself noted the difference:

> "Right now, I'm wondering: Have I done or said anything amiss? You see, at this moment everything looks clear to me, but what happened just before? That's what worries me. It's like waking from a dream; I just don't remember."

Although HM could remember his childhood, he was unable to form new memories after the surgery. If the same clinician interviewed him two days in a row, HM would react as if it was the first time they had met. Lore has it that HM was once served three plates of food successively. After the third he stopped midway, simply noting that he was no longer hungry. His inability to remember new facts after his surgery was so severe that it revolutionized thinking about how memory works.

But not every new thing encountered was forgotten. Researchers soon discovered that there were things that HM could learn. For example, when he placed his finger on a dot and then followed it on a rotating disk (a record for those of us who remember life before the late 1980s) his improvement over successive sessions was indistinguishable from neurologically intact controls. Over time he could keep his finger on the rotating dot for longer periods of time. Unfortunately, he had no sense of his improvement. He consistently claimed to be doing this task for the very first time. These findings led researchers to conclude that not all forms of memory are the same. Specifically, learning *about* things relies on different neural systems than learning *to do* new things. The difference between declarative (about) and procedural memory (doing) has become one of the building blocks of cognitive psychology.

The distinction between these two forms of memory also helps to explain how experts and novices differ. During the learning phases of a new activity, novices spend considerable cognitive resources consciously engaging and monitoring their activity. Consciously thinking about a skill allows them to make changes to any particular component that might be deficient. As novices engage in extended practice, they begin to consolidate these pieces into a whole. Hence, over time extended practice leads to a proceduralization of these processes (Fitts & Posner, 1967; Keele, 1968; Kimble & Perlmuter, 1970; Proctor & Dutta, 1995).

THE BRAIN BASES OF EXPERTISE

So far we have discussed how expertise seems to arise as certain aspects of a task move from declarative to procedural memory. Furthermore, it was found that HM could learn new motor skills but not learn anything about them. Taken together these two strands of research would suggest that higher expertise would be associated with the use of procedural memory brain circuits and less expertise with the use of declarative memory brain circuits.

The ways in which declarative and procedural memory systems are involved in learning have been explored elegantly by F. Gregory Ashby of the University of California, Santa Barbara (Ashby & Crossley, 2012). In a series of studies, work in his laboratory has looked at the ways in which adults learn to categorize stimuli. In these experiments, learning for categories that have rules that are easy to verbalize is compared with learning for categories that have rules that are not easy to verbalize. This allows him to compare rules that use declarative memory with those that use procedural memory.

One way to think about this is to consider how people learn to locate items in a grocery store. There are some basic rules you can learn rather quickly. Generally, the flour is near the sugar. The vegetables and fruits are on one side, whereas the meat and poultry are on another side. The freezer section is not too far away. The baked goods sit in an entirely separate section. So we can think of a supermarket as consisting of different hubs. This relies on the declarative memory system, since one can easily tell people about these things.

Finding items in a supermarket would be easy if such a simple spatial catalog existed. The problem is that each supermarket has its idiosyncrasies. Invariably, a person may end up searching for peanut butter, organic breakfast flakes, or apple sauce. They are in certain sections but trying to find them can be very difficult. What ends up happening is that individuals wander aimlessly through the aisles until they finally find an employee who can tell them where it is. Then the employee will rattle off the aisle number as if it were common sense. And at that point, a customer may think what I often think: Why on earth is it there?

The distinction between these clear hubs that can be talked about out loud and those specialty items that do not have the same location in every store captures the essence of declarative and procedural memory. Interestingly, the real world requires a mix of these abilities. Ashby and his colleagues have designed laboratory tasks where participants are trained in mapping visual stimuli on a computer screen with rules that can be stated out loud such as sets of right leaning, /// or left leaning bars \\\. In another condition, participants had to learn visual categories that had no explicit rules such as a set of icons in category A (✦▲❍➲✻) or category B (♣⇨◆✈❡). Coming up with a clear rule for these two categories is not straightforward, and hence people tend to simply learn through consistent practice.

Interestingly, the nature of brain activity shows changes in the particular brain systems that are used over time. Because both of these rules involve mapping a visual category with a motor response, they are heavily regulated by the basal ganglia, a set of areas that are highly interconnected with motor cortex and premotor planning areas in the frontal lobes. Initially, category rule learning involves two separate brain systems depending on the type of rule being learned. For categories that relied on declarative memory, like the hubs in your local supermarket, there was increased activity in the prefrontal cortex, the head of the caudate nucleus, and the hippocampus. Hence, the ability to verbally state what is being learned is moderated by the very area that HM lost along with two areas that are highly interconnected in the cortex (prefrontal) and the basal ganglia (caudate). These two latter areas are known to be involved in cognitive control (Provost, Petrides, & Monchi, 2010), a topic I will cover in more depth in chapters 6 and 7. Category learning that depends on procedural memory, like those esoteric items that seem to rely on the idiosyncratic aspects of each supermarket, shows neural activity in the putamen, a brain area that is known to be involved in motor control. Hence, each type of learning involves different brain circuits that are thought to mediate different forms of memory.

By the 20th session, however, performance in either category no longer differed. Thus practice leads to a change in the way in which the task is performed. The interesting part is that in later sessions performance was linked to activity in premotor cortex. Eventually, with enough practice in any particular supermarket both the items that have clear hubs and the specialty items that do not will be handled with the same neural mechanism. A person will simply go to that aisle and get the item. To do this requires motor planning, a specialty of the premotor cortex, but not much else.

The work that has taken place in Ashby's laboratory is interesting with regard to the expertise literature. Specifically, it suggests that training in categories that rely on procedural or declarative knowledge eventually lead to the same brain result. People are able to make decisions using simple motor planning. Hence, gaining expertise on a task over time leads to task performance being linked to motor planning systems. Expertise can be seen as a form of neural efficiency.

The notion that neural efficiency differentiates those with higher expertise in a domain has received attention in the literature. For the most part, there is a lot of evidence that experts show less neural activity relative to novices on a variety of intelligence tasks that tap into both spatial processing and into working memory, which involves keeping several things in mind temporarily (Neubauer & Fink, 2009). For easy to medium tasks, there is generally more activity for subjects who score lower on psychometric tests of intelligence, suggesting that their brains are less efficient and have to work harder. As noted earlier in the work of Ericsson, expertise seems to override the effect of overall IQ both behaviorally and neurally. For experts in a domain, there is no longer a correlation between the amount of frontal lobe activity and performance on a task. Rather experts in a domain come to rely on the parietal lobe or the hippocampus to perform tasks in their domain of expertise. That is, expertise leads to the use of posterior brain areas that are more directly tied to the processing of a task.

EXPERTISE AND LANGUAGE PROFICIENCY

Despite being a point of major consideration in the bilingual literature, the construct of proficiency gets a lot less mention in the monolingual literature. However, we all know that people speak their first language with differing amounts of ability. Even in the highest groups of college bound high school seniors, there is variation in their English ability. Not even extensive test preparation can eliminate the variation in scores of high school seniors. Even among those who score very high on the test, some may be better at writing poetry or at debate. All of these skills involve a specialized use of language. And practice in all these domains leads to considerable improvement. So what are the domains of language that also reflect differences in proficiency?

In the mid-1990s, groups of researchers began to look at functional brain differences that differentiate monolingual groups of subjects who differ in language proficiency. One of the first studies to consider how proficiency would impact brain activity in monolinguals was conducted by King and Kutas (1995), who looked at the EEG waves of a group of participants while processing two different types of sentences:

1. The reporter who the senator attacked discovered the error.
2. The reporter who harshly attacked the senator discovered the error.

After presenting sentences like this participants were asked to answer questions (i.e., Who discovered the error?). If you are like most English speakers you will find sentence 1 more difficult than sentence 2, even though the meanings of the two are very similar. Sentence 2 uses a more conventional layout and is much clearer. The reporter attacked the senator and also discovered the error. Sentence 1, on the other hand, is more difficult to understand. This is partly due to the fact that each verb has different points of emphasis. In fact, you could rephrase the sentence entirely and make it easier. That is, the senator was attacking the reporter who discovered the error. The problem is that the sentence then loses its emphasis that the reporter discovered the error. In order to keep the focus on the reporter the sentence requires the reader to keep two ideas in mind. The higher cognitive processing load required to keep this information in mind results in the electrophysiological signatures associated with each type of sentence. King and Kutas were also interested in understanding the role that language proficiency plays in processing simple and complex sentences. In order to differentiate their subjects, King and Kutas used a measure of verbal working memory originally developed in a different laboratory in the early 1980s.

In order to measure working memory, Meredyth Daneman and Patricia Carpenter developed a task in which participants were asked to to read a set of sentences out loud while remembering the last word of each one (Daneman & Carpenter, 1980). You could try this by reading the first three sentences of this paragraph out loud while remembering the last word of each sentence. The number of words that participants could remember is indicative of their working memory capacity. Work using this reading span task has captured considerable differences in people's ability to remember words while reading (Daneman, 1982, 1991; Daneman & Carpenter, 1980, 1983; Daneman & Green, 1986; Daneman & Merikle, 1996; Daneman, Nemeth, Stainton, & Huelsmann, 1995). The notion is that increased proficiency in a language would result in a larger reading span.

In testing participants using ERPs, King and Kutas (1995) found that poor and good comprehenders differed in the late anterior negativity (LAN), a negative going wave that appears between 250–600 milliseconds over the left central electrodes. It has been shown to appear as subjects repair sentences with grammatical errors. The results revealed that poor comprehension was associated with a larger LAN.

The interesting aspect of King and Kutas's data is that there were no grammatical errors (King & Kutas, 1995). Hence, the increased LAN that was found for grammatically correct sentences suggests a new conceptualization of repair, since both types of sentences are grammatically correct. What this specifically suggests is that each sentence requires a different amount of processing effort in order to interpret. Hence, it is as if participants were reordering the sentence so that they could interpret it.

Differences in verbal working memory proficiency did not uniformly predict differences in the brain waves observed. For the easier sentences in which the agent of a sentence remained the same (i.e., The reporter who attacked the senator…) there was no difference across participants who differed in proficiency. However, for sentences where the agent changed (The reporter who the senator attacked…), groups that differed in language proficiency showed very different neural signatures. When poor comprehenders encountered the verb inside the clause (i.e., who the senator **attacked**) they showed a larger LAN than the good comprehenders. Finally, poor comprehenders also showed a large slow positivity that persisted for the entire back end of the sentence. It is as if the brain of poor comprehenders was so busy trying to make sense of these sentences that it kept futilely firing up its neural generators to no avail. Naturally, this implies that those with higher language proficiency where able to process these sentences with much less cognitive effort. In short, as with nonverbal tasks, proficiency was efficiency.

A natural extension of this initial study would be to see if proficiency also applies to different amounts of neural activity across monolinguals who differ in language proficiency. To test this question, Christine Weber-Fox and colleagues asked a group of college students to read sentences that contained nonsense endings (The student looked at her watch to check the rain) and those that did not (The boy put the worm on the hook). The results revealed differences in the time at which the different brain waves appeared in each group. High proficiency speakers were better able to process the words, figure out their meaning, and integrate them into the existing sentence structure. One way to visualize this is to think of a set of waves that flow across the scalp. The waves of the high comprehenders moved faster and more efficiently than those of the low comprehenders.

Finally, one more question is whether the same effects also appear in paradigms that require auditory processing of language. To test this, Erik Pakulak (Pakulak & Neville, 2010), a graduate student at the time working with Helen Neville, asked a group of monolingual English speakers to listen to sentences that either did or did not have grammatical errors. For example, participants heard sentences like:

1. Timmy can ride the horse at his farm.
2. Timmy can ride the horse at my **his** farm.

The bolded word indicates where the second sentence goes wrong. Pakulak and colleagues found results that parallel those discussed above. Namely, high proficiency speakers of English had much more focused activity both in time and space for the LAN and the P600. Not only were their brain waves smaller and more intense but also they are shorter. It is as if the brain of high proficient users was tuned to language so finely that it could register the errors and repair them quickly and efficiently. Again, proficiency is efficiency.

So far we have discussed the nature of language proficiency with regard to the electrical brain waves generated by the response to both easy and difficult sentences. The question this naturally leads to is whether fMRI, which indirectly measures the firing of neurons, can tell us more about that brain regions that differentiate high and low proficiency individuals.

Work by Chantal Praat from the University of Washington along with Marcel Just from Carnegie-Mellon University, has revealed what brain areas differentiate those with higher and lower proficiency (Prat & Just, 2011). In one study with 84 readers, they found that those with lower vocabularies showed more brain activity than those with higher vocabularies. The areas of increased activity included a network of areas in the frontal and temporal lobes that are involved in the control and use of higher-level language. In the left hemisphere, there was increased activity in the inferior occipital and temporal cortex, an area known to be involved in the recognition of visual forms and is known to be involved in word recognition. Speakers with lower proficiency generally engage a larger network of areas in order to read sentences. Hence, the brains of those with higher language proficiency exhibit greater neural efficiency.

PROFICIENCY AND EFFICIENCY AS AN EXTENSION OF PITRES'S RULE

In this chapter we considered the importance of proficiency and expertise as an important explanatory concept across multiple domains. As Pitres deduced many years ago, familiarity plays a strong role in the processing of language. The rub is that AoA plays a role as well. So which is the most important factor? Like many questions in science there is no simple either/or kind of answer. But what is clear is that improvement in a skill occurs over time through hours of exposure. This can be seen in both the general expertise literature as well as in the fact that children from families with different socioeconomic strata are exposed to speech that differs in both quantity and quality. However, simple exposure is not enough. Another crucial component is that this exposure be coupled with feedback in order to fine tune the cognitive and linguistic system. As the system becomes more and more specialized it leads to a form of cognitive economy. This economy allows faster and more efficient encoding with substantially less effort. With this in hand, the expert whether pondering his or her next move or listening attentively to a lively debate is able to do more with less.

5

PRACTICE WHAT YOU SPEAK

IN PITRES'S ORIGINAL view, familiarity with a language was the most important driver of recovery of function. Although the importance of this can be seen in the cases Pitres originally presented many years ago, the importance of familiarity can also be seen in more recent cases. In 1996, just as I was transitioning between graduate school and a postdoc at the University of California (UC), San Diego, Estrella Gomez-Tortosa and colleagues published a study that provides an interesting twist on Pitres's rule (Gomez-Tortosa, Martin, Gaviria, Charbel, & Ausman, 1995). In this article, Gomez-Tortosa et al. report the case of a 22-year-old woman who had moved to the United States from Bolivia at age 10. At age 19 the patient began to exhibit naming difficulties along with seizures. A brain CT scan revealed brain damage that affected both the left temporal and parietal lobe, an area known to be involved in the processing of auditory language. An operation was performed to remove a venous malformation in the area.

The researchers had the foresight to look at language proficiency in the patient before and after surgery. A series of standardized tests showed some deficits in her ability to attend to sounds and in her verbal memory. In addition, the Boston Naming Test, long used to test patients' ability to name pictures, was also administered. Before the operation, half of the items were administered in Spanish only, yielding a score of 19 out of 30. Administration of the test after the operation yielded a score of 32 out of 60 in Spanish and 44 out of 60 in English. The patient's report of having difficulty with her speech in Spanish was reaffirmed by family members, who confirmed that she often appeared to say words that had no meaning or to confuse words and say the wrong one.

Gomez-Tortosa and colleagues interpreted the findings as supporting the view that different brain areas are devoted to each language. That is, the greater decrement in Spanish, the native language, relative to English, the second language, is evidence that damage to the temporal and parietal lobes affect the more familiar language, Spanish. However, there are two caveats that we should consider with regard to this hypothesis. First, it is difficult to determine exactly which language was more dominant before the operation. On the surface it appears that Spanish showed a larger decrement since there was a drop from 19/30 to 32/60. To conclude this we would have to assume that scores on the picture vocabulary test are valid whether the whole test is given or half the test is given. Second, we would have to assume that Spanish is the most familiar language and that English is less familiar. However, there is no data from a comparison group of bilinguals without brain damage. As we'll see in this chapter, the relationship between proficiency and age of acquisition (AoA) of a language breaks down in some bilingual populations. For the remainder of this chapter, we will explore the disjointed nature of language in bilinguals and the impact this has on the development of proficiency across age.

ON LANGUAGE LOSS

When I was two years old my mother was shopping around for in-home daycare and was referred to Mrs. Neshati, a Persian mother who took care of children along with her daughter. I seemed very happy there and my mother thought nothing of leaving me in a Persian-speaking home. After a few months she noticed something peculiar. Mrs. Neshati was speaking Persian—Farsi—as she always did with her daughter at home. But the response did not come from her daughter. It came from me!

My mom was surprised as Persian sounds slowly began spreading into our household. The arrival of these new words was accompanied by growing frustration on my part when my mother couldn't understand them. Both my parents spoke Spanish to me. But I was exposed to English through neighbors, television, and interactions with friends. So my mom thought I was going through a period of English-Spanish bilingual adjustment. She was confident that eventually I would sort it all out and both her frustration and mine would slowly fade away.

After about a year my mom felt it was better that I move to a more formal preschool. When we left Mrs. Neshati behind, the parade of Persian sounds that had come to live in our house began to retreat and within a few months they were gone. And no, I don't speak or understand Persian today. This language is seemingly gone forever.

This loss makes sense to most people. After all, I was not exposed to a foreign language for that long. As I grew older, my dominant languages would literally overwrite the previously learned language. Previous research has also shown how exposure to a second language early in life can lead to eventually having better proficiency in that language (Austin, 2007). The fact that I learned and lost Persian at an early age reveals that both AoA and length of exposure matter in determining the ultimate attainment in any particular language. It also suggests that early in language learning there is an apparent vulnerability in any language that is learned. Consequently, every child could potentially lose a language despite having very good proficiency in that language. The question that remains is whether a first language, one that is used on a regular basis, can be lost.

CATASTROPHIC LOSS

The most dramatic form of loss has been documented by Christophe Pallier, an expert in the brain bases of language (Pallier et al., 2003). In a now famous study, Pallier asked a group of adult Koreans who had been adopted by French families between the ages of 3 and 8 to perform a series of tasks inside and outside of the magnetic resonance imaging (MRI) scanner. Outside of the scanner, subjects had to rate whether a set of auditory sentences in many different languages was presented in Korean or not. In addition they had to match a written French word to the correct Korean translation. In the scanner, participants listened to sentences in one of four languages (Korean, French, Polish, and Japanese). Results from outside the scanner revealed identical performance when comparing French monolinguals with the adoptees. Furthermore, the brain scans showed almost identical patterns of activity in both groups. In short, it is as if the native language had been entirely forgotten by adulthood. What is particularly stunning is that the adoptees had moved to France during childhood well after the first language had been established. Yet all of them showed the same pattern of performance: no observable trace of their first language.

One potential criticism of this first study is that the tests used might not have been sensitive enough to detect the residual language knowledge that was present in a population that had been cut off from its first language. To extend this further, Ventureyra asked the same group of adoptees to listen to Korean speech sounds that are particularly difficult for nonnative speakers to recognize (Ventureyra, Pallier, & Yoo, 2004). This was done to test the possibility that early exposure to Korean would give the adoptees sensitivity to speech sounds they had heard early in life. Since they were exposed to Korean for at least three years, they should have heard and used these sounds extensively, thus retaining some sensitivity to their native language. The results were the same as in the imaging study I discussed above. Both groups were unable to recognize these sounds. The results revealed, once again, Korean adoptees were virtually identical to the French speakers.

CAN A LOST LANGUAGE BE FOUND?

The results from Pallier's and Venturerya's studies suggest that complete immersion in a second language early in life might lead to complete loss of the first language. But let me pause here for a second and go back to the question that Pallier asked: "Can a first language be lost?" Notice that "lost" does not mean destroyed or annihilated. Things that are lost can be found. In short, this loss may not actually be permanent. Some might reason that the first language was not gone, it was simply not retrievable after years of being immersed in their second language (Bjork & Benjamin, 2011). Cases that appear to involve forgetting can occur quite frequently in everyday life. I recall running into an old high school acquaintance at my 10th-year reunion. He recognized me immediately. But my life had become socially disjointed after high school. I had spent two years at Berkeley, two years in Brazil, and then two years back at Berkeley. After this I had gone off to graduate school at the University of California (UC), San Diego, where I had been a teaching assistant for several classes. In addition, the Center for Research in Language had parades of visitors streaming in from the entire world. By the time I made it to my high school reunion, I was suffering from person memory fatigue. I could remember those whom I had seen frequently very easily. But I stood frozen as my peer spoke to me in a very enthusiastic tone. It took quite a while to place this particular individual. I hesitated and slowly the faint memory began to come back. If I were to see him again, I would have the benefit of having seen him a few years after having completed high school. I remember him better today than I did when I first saw him back in 1995. This is due to my original memory trace being rekindled by seeing him once again (and also by writing about it here). Had I gone several decades without seeing him, I might have forgotten all together. Or at the very least I would have had an even more difficult time remembering who he was.

The interesting part of memory is that there is the ability to relearn things. A similar question exists in the language literature with regard to foreign language adoptees. Can these participants actually relearn their lost language faster than native speakers? This would help to elucidate the ways in which language is subject to the same laws that rule memory in general.

Examples of relearning in language have been found by another group of researchers. In chapter 3, we discussed studies by Terry Au and colleagues (2002) showing that adults exposed to a language in childhood were able to relearn certain aspects of that language. Specifically, those who had listened to a language in childhood showed maintenance of the phonological aspects of that language. Unlike the adoptees tested by Pallier and colleagues, these children had remained in their original families in the United States. Although they did not speak their maternal language, it is difficult to ascertain the precise amount of exposure to their native language. Thus it is unclear whether this group's results are directly comparable with the results obtained by Pallier and colleagues in France.

More recently, Oh et al. recruited a different group of Korean adoptees who were enrolled in one semester of Korean at a university (Oh, Au, & Jun, 2010). Eleven of the adoptees had been adopted before age one, and one was adopted at age three. The adoptees had minimal exposure to Korean, mostly through camps that brought together adoptees from the same culture to expose them to their home culture and, to a minimal extent, to the language. Like the work by Pallier and Ventureyra with Korean adoptees in France, there was no overall advantage of adoptees over monolinguals in terms of accuracy. However, upon closer examination Korean adoptees showed an advantage on certain speech sound distinctions that exist in Korean but not in English. In short, it appears that Korean adoptees do seem to retain some basic knowledge of the speech sounds used in their forgotten native language.

Similar results have been revealed in the work of Kenneth Hyltenstam and colleagues, who have found that adoptees show residual sensitivity to speech sounds above those seen in nonnative speakers (Hyltenstam, Bylund, Abrahamsson, & Park, 2009). In addition there is evidence in these data that adoptees from Latin America retain some nonnative features of speech. From these results, Hyltenstam concludes that children can become close to native-like after extreme foreign language immersion during the first decade of life. However, the ability to achieve this appears to be related to when the immersion occurs, with earlier immersion favoring more native-like pronunciation and grammatical processing.

Taken together these results lend support to Pitres's original proposal: a recently used language might actually be more engrained in memory and hence more resistant to loss. Specifically, early-learned languages are extremely susceptible to interference if the language of the environment changes. The tricky part in all of this is that what is lost or retained depends on the particular language function we are investigating. The examples we have encountered so far suggest that phonology seems to be particularly resistant to loss.

PLASTICITY AND LANGUAGE ACQUISITION

The fact that a first language can be replaced by a second language lends support to Pitres's view of language proficiency to some extent. It suggests that language ability is highly sensitive to environmental input and requires hours of deliberate practice before becoming solidified. There are two possible ways that this could happen. The first would be that children learn a protolanguage, a very stripped down version of adult language. This stripped down version is so incomplete that it can be "erased" during development. This would suggest that the environment (i.e., the host country for immigrants) would be driving language learning. A second, more reasonable assertion would be that at least some of what a child learns early in life applies to language learning in general. Learning basic steps in Korean would help in learning French or English.

For most adults, this makes no sense. I once mentioned this to my cognitive psychology class. A young man sat befuddled and wondered out loud how I could think his first language helped him to learn English. For the adult or adolescent language learner, the first language appears to be a hindrance to second language learning. It makes them have accents or stumble to come up with the right words. Based on Pallier's work, however, there appears to be an advantage to learning a first language even if it does not match the language that is learned later in life. If this were true, then we should find evidence that lack of exposure to a language impairs future language acquisition. To prove this point it would be helpful to look at cases of language deprivation.

I noted earlier two cases of language deprivation that became famous: Victor of Aveyron and Genie (Candland, 1993; Curtiss et al., 1974). Both Victor and Genie ended up as social outcasts who were never able to blend in with human society. So language deprivation has severe effects on the ultimate ability of a human to adapt to society. The main question is whether this is due to language deprivation, social deprivation, or both.

To test whether it is language deprivation, per se, that is the main cause of problems with language learning, it is important to find a population that is socially engaged even though it is cut off from language. Unlike hearing children, those who are born congenitally deaf are not necessarily exposed to language from birth. If their hearing is not corrected, or if they are not exposed to sign language at an early age, these children will be effectively cut off from the early language input that hearing children experience. By studying this population we can get a glimpse into a mind without early language exposure that nonetheless is exposed to social interactions with caregivers.

Back in the 1970s, fewer than 10 percent of deaf children were born to families with an American Sign Language (ASL) speaker. Many of these children were effectively cut off from language during the early stages of development and were not exposed to it until well after their first year, with many of them not learning language until they entered school. Earlier recognition of deafness has led to a reduction of these cases. However, there are still children who are effectively deprived of linguistic input at a young age. The fact that deaf children vary in the age at which they receive language exposure allows one to ask the this question in a much more systematic manner. Specifically, we can ask how age of first language acquisition affects language development.

This issue has been addressed by Rachel Mayberry and colleagues in a series of studies (Boudreault & Mayberry, 2006; Mayberry, 2007; Mayberry & Lock, 2003; Mayberry, Lock, & Kazmi, 2002). In the most publicized study, Mayberry asked a group of deaf and hearing participants to decide if a set of sentences were grammatically correct.[1] Half of the items were grammatically correct sentences, such as "I want you to go to the store now." The other half were incorrect sentences, such as "I want you will go to the store now." Similar types of sentences were also constructed in American Sign Language (ASL). All of the subjects who were exposed to either ASL or a spoken language in childhood outperformed those who had no early language exposure. There was no additional benefit when early exposure matched the language of testing. In other words, deaf subjects performed just as well on the ASL test whether they were exposed to ASL or English as infants. On the English version of the test there was also no effect of native language on the results. These findings suggest that early language exposure clearly matters. But it doesn't matter in the way one might think. Rather than affecting the development of any particular language the benefits spread from one language to another.

The results from studies with children who are exposed to multiple languages early in life offer an interesting set of findings that qualify the language acquisition process. Young children can learn a second language very well early in life. Hence, the language talent that children show early in life appears in their ability to acquire additional languages. However, in some cases this ability actually results in severe language loss. Whereas work with monolingual children suggests that children's skills become specialized very quickly, work with bilingual children and adoptees suggests that the ability to learn a second language remains open for quite some time. In fact, there is enough flexibility in the system that one can literally forget a first language at least during the first decade of life. The loss of a first language occurs in tandem with the acquisition of the second language that rivals that seen in monolinguals. Taken together these results suggest a much more staggered type of development than the one supported by models of single language development.

RESTRUCTURING PROFICIENCY

In the last chapter we considered the notion that proficiency in a language is related to the number of hours spent interacting with others and the quality of that interaction. Over time the ability to chunk complex language computations into a single step allows those with higher proficiency to keep more things in their minds temporarily. I also noted last chapter that language proficiency is an interesting extension of expertise across many domains. Specifically, language expertise is associated with automaticity, in which declarative and procedural memory migrate to premotor planning. Expertise leads to an implicit slicing up of actions into parts, and talking about them becomes increasingly difficult. This leads to a very clear difference between novices and experts.

In bilinguals proficiency has also come to be seen as involving differences in language experience and expertise. As we saw in the language loss literature, bilingualism brings up another interesting facet that is not often discussed in the expertise literature: the effect of prior experience on current learning. We can try a sports analogy to illustrate this point. For example, take a person who is trying to learn to play tennis. This person might have played softball and soccer as a child. She might borrow the mechanics of throwing to serve, the movement of catching fly balls to track overhead smashes, and the batting motion to hit forehands. Soccer would help with footwork, spacing, and weight transfer. Hence, a person does not arrive at a sport starting truly from zero. In reality, a new sport can build on skills that might have already been learned in other sports. Seen from this perspective, learning a new skill involves more than just the time spent on a particular task. It also involves the time spent building up a set of subskills that can then be reconfigured into a new skill. In short, any newly learned aptitude can be viewed as a patchwork of previously learned subaptitudes.

In extending this sport analogy a bit further, we can think of a deeper more complex aspect of learning. During the early years of sports, there might be considerable time spent on mechanics. A budding tennis player will spend time figuring out how to tilt her shoulder and arrange the path of her racket in order to optimize the serve. A similar problem arises in soccer. At first children learn to kick facing forward. But in time they are taught to turn their torsos and hips in order to kick harder. Thus early childhood involves learning basic mechanical aspects of a sport.

As children get older, sports becomes more than mechanics. It will come to involve strategy. High school athletes will be given "playbooks" and watch videos. Others will scout their opponents to see their strengths and weaknesses in order to better develop a strategic plan. This type of thinking relies on a set of higher-level cognitive skills. Whereas first graders spend all their time chasing the player who has the ball, by high school you can see a soccer team nicely spaced out, like the idols they watch during the men's and women's world cup that is played every four years.

The notion of language building up different skills is not frequently discussed in the language acquisition literature. However, the fact that learning involves taking a set of existing skills and reconfiguring them into a new skill is embraced more easily in the second language learning literature. Because each language is identified as an entity in and of itself, bilingual researchers have been more likely to consider how a second language builds on the first. This disjointed nature of language processing can be seen in the fact that a second language can replace a first when learned in childhood. At the same time adults show residual sensitivity of speech sounds that are unique to the first language.

The distinction between shared and/or separate systems and its implications for language proficiency has also made its way into the bilingual psycholinguistic literature (Kroll & Tokowicz, 2005; Kroll, Van Hell, Tokowicz, & Green, 2010). In recent years, researchers have come to agree that bilinguals, for the most part, have a single set of concepts that are linked to separate words in each language (Altarriba, 1992; French & Jacquet, 2004; Kroll & de Groot, 1997, 2005; Kroll & Tokowicz, 2005; Kroll, Tokowicz, & Nicol, 2001; Potter, So, von Eckardt, & Feldman, 1984; Schreuder & Weltens, 1993; Sholl, Sankaranarayanan, & Kroll, 1995). Hence, the concepts on which a bilingual's two languages are built rely on roughly the same world knowledge.

Despite relying on the same world knowledge, bilinguals still show differences in their ability to link words to concepts in each language. The fact that the link to a concept depends on proficiency is most clear in the revised hierarchical model of Kroll and colleagues (Kroll et al., 2010). In this model, as depicted in Figure 5.1, each language has its own separate word store and a conceptual store. There is a set of two-way connections between the first language (L1) word store and the second language (L2) word store. In addition, each conceptual store and each lexicon also has a set of connections.

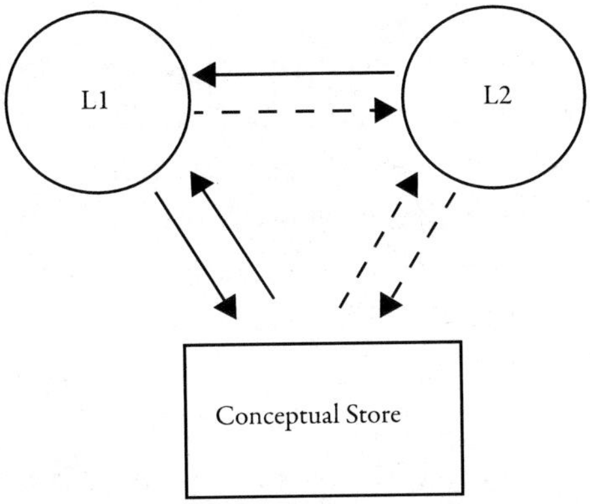

FIGURE 5.1 The revised hierarchical model (Adapted from Kroll, J. F., and Stewart, E. Category interference in translation and picture naming: Evidence for asymmetric connections between bilingual memory representations. *Journal of Memory and Language, 33*, 149–174. Copyright (1994), with permission from Elsevier).

The initial stages of second language learning involve using the first language as a crutch. One way to think about language proficiency is to think of it as relating to the strength of these connections. During initial learning the connections between the two word stores are asymmetrical in strength. The second language is more strongly linked to the first language than the first is to the second. That is, a second language is parasitic on the first (MacWhinney, 2004).

Early in second language learning, adults often report that they need to translate a language in order to understand it (Kroll & Stewart, 1994). Because the links between the first language and concepts are stronger, it is more efficient to simply translate the word and then access the concepts. A series of studies have confirmed this model's predictions. Access to the meaning of a word is faster in a first language compared to the second language. And it is faster to translate into the first language than vice versa. As proficiency improves, the second language begins to build connections to concepts. Despite improvement in proficiency, the asymmetry in the strength of the connections remains for the majority of late second language learners (A. Hernandez, Li, & MacWhinney 2005).

The fact that concepts are shared and words are not leads to an intriguing possibility. Specifically, it is possible that parts of language that reinforce general conceptual knowledge would actually be strongly tied together. Aspects of language that were based more on words would be less connected. In other words, there is nothing about the letters or sounds that make up "*coche*" that helps with the recognition of "*bil*," its Turkish translation. Hence, the disconnection between languages is greater the closer we get to the actual form of words.

THE DICHOTIC NATURE OF LANGUAGE

The fact that there are differences between words and concepts brings up an interesting and important distinction that has been picked up in a seemingly unrelated literature. In 2003 I received an e-mail from Lynette Austin, who was a first-year graduate student in educational psychology at the University of Houston. Lynette was interested in taking my course on the neural bases of language. It quickly became clear that Lynette's interests and mine coincided in a very quirky kind of way.

For a long time work with early Spanish-English bilinguals had found a reversal in language dominance[2] (Gollan, Montoya, Cera, & Sandoval, 2008; Gollan, Montoya, Fennema-Notestine, & Morris, 2005; Gollan, Montoya, & Werner, 2002; A. E. Hernandez et al., 1996; A. E. Hernandez & Reyes, 2002; Kohnert & Bates, 2002; Kohnert, Bates, & Hernandez, 1999). The interesting thing about this population was that it seemed to show a completely reversed effect relative to that observed in adult second language learners. Specifically, early bilinguals who had learned Spanish first were coming into the lab as adults with higher proficiency in English. They could name more pictures, were faster at reading words, and had better literacy skills in English.

The fact that the age of learning can become uncoupled from proficiency in a language was also present in Lynette's mind as she came back to graduate school. Lynette had learned Spanish in college and spent seven years as an adult living in Argentina, where she had learned Spanish so well that she was able to give professional presentations entirely in Spanish. But the real eye opener for her was the effect that this had on her children. Her oldest child, Alison, started grade school and learned Spanish very well. Her son, Daniel, who was born in Argentina, left the country speaking two- to three-word combinations in Spanish. However, over the ensuing years he lost his ability to speak Spanish but retained a really good accent and could even trill his *r*'s, a task that is very difficult for nonnative speakers. He often answered the phone and spoke so well that he was quickly engaged in longer one-sided conversations before he was rescued by his parents or older sister who spoke better Spanish. This disjointed experience had primed Lynette for one of the biggest distinctions in the bilingual literature. Namely, that someone could sound like a native even though he or she had little knowledge of the language.

Armed with her personal experience in Argentina and her professional experience as a bilingual speech-language pathologist, Lynette arrived with her research question very well delineated. What she noted was that there were faculty members who were very proficient and highly educated but spoke English with an accent due to their later acquisition of this language. On the other hand, there were many native speakers whose speech would be rated as better than the nonnative faculty members who had heavy accents. Yet, some of these native speakers might actually score lower on a standardized language proficiency test. Lynette was referring to two different ways in which language proficiency might be evaluated. One involved basic everyday language use. The second involved more academic language.

The distinction between academic and everyday language was first brought to prominence by James Cummins in 1980. Cummins found himself in the middle of a debate on the nature of language proficiency in the late 1970s. The notion that proficiency may be a single unified cognitive function was championed by the work of Oller (1978, 1979), who suggested that language proficiency was highly correlated with overall IQ and literacy levels. In his view there was little room for variation in specific skills within language. On the other hand, Eduardo Hernandez-Chavez, a linguist with an interest in bilingual language use, suggested that language could be divided into many different parts (Hernandez-Chavez, Burt, & Dulay, 1978). These included what we typically refer to as syntax, semantics, and phonology.

In addition, it includes written and spoken language as well as sociolinguistic variables that might influence how language is used depending on the person or the context in which the conversation takes place. In short, there were two opposing views: one in which language was directly related to IQ and another in which language was composed of a variety of subskills.

Jim Cummins stepped into this debate in 1980 and suggested that both views were valid. There were aspects of language that are typically associated with academic performance. In general these language abilities are highly correlated with IQ or standardized verbal abilities tests. However, there were also language abilities that do not correlate strongly with academic language. For example, Cummins noted that standardized measures of language competence were weakly related to spontaneous language use (Wells, 1979). This view is also echoed in more recent work which has found that tests of standardized vocabulary correlate weakly with the number of different words used and do not correlate at all with the total number of words used in spontaneous speech samples (Ukrainetz & Blomquist, 2002). The fact that spontaneous speech samples do not correlate strongly with standardized language assessments suggests that language skills based on the ability to communicate in everyday life may represent a separate type of language competency. This competency includes accent, oral fluency, and sociolinguistic competence.

To account for these two distinctions, Cummins came up with the terms "cognitive and language proficiency" (CALP) and "basic interpersonal communication skills" (BICS) to highlight these different types of language proficiency. Cummins proposed that these two skill sets, although interrelated, were in many ways quite dissociable. As Lynette had noted, BICS and CALP may come apart. In most monolinguals, where a single language system was built from the bottom up, the difference between these two forms of language ability is not always clear. When language learning is disjointed, as had occurred with Lynnette's children in Argentina, these two aspects of language can come apart. A similar effect appears in the adoptee literature we discussed earlier, where there is some retention of phonological aspects in the first language with virtually no ability to even recognize words. Hence, bilingual speakers show quite revealing differences between BICS and CALP.

The notion of BICS and CALP also fit nicely with the predictions that the revised hierarchical model makes for late learners of a second language. In Kroll's model each word store is linked to a conceptual store. Initially, these links are stronger in the first language. However, over time the links grow stronger between the second language and the conceptual store. Eventually, with enough practice, late learners can access concepts directly without relying on translation. In the case of the adult classroom learner, there is a more cognitive approach taken to learning, since words and concepts are linked together as second language proficiency improves.

BICS and CALP, which capture the difference between everyday and academic language, respectively, serve as an interesting way to break bilingual language processing apart. However, the differences between BICS and CALP also allude to a potential difference in the factors that might be influenced by these two forms of language use. Since BICS is more involved in everyday language, it might be related to language use early in life. CALP on the other hand may be more related to the use of language in a school setting. How this plays out in the bilingual literature is of particular interest.

WHAT IS PREDICTED BY LANGUAGE PROFICIENCY?

The fact that proficiency involves connections between words and concepts that are involved in the use of academic language would lead to a link between semantic processing and language proficiency. This link can also be seen in two studies by Weber-Fox and Neville (1996) and by Isabel Wartenburger and colleagues (Wartenburger et al., 2003), which we discussed earlier. In these studies, participants were asked to read sentences that contained grammatical and semantic violations. Whereas grammatical violations were sensitive to when a language was learned, semantic violations were related to proficiency in a particular language. Thus it appears that proficiency in a language is tied to processing the meaning of a sentence.

Weber-Fox and Neville's study measured EEG waves as subjects read sentences that contained semantic violations such as "The scientist criticized Max's **event** of the theorem" and sentences that had no violation. The congruent sentences contained a word such as "explanation" instead of "event." As noted earlier, the semantic error led to an increase in the N400, a negative going wave that peaks at 400 milliseconds after viewing the word. The question that Weber-Fox and Neville sought to address was to what degree factors that are important for second language processing such as AoA or proficiency would play a role in the size of this negative going wave. The interesting part of their findings is that changes in the N400 were only noted in participants who learned English after age 11, a point at which their proficiency dropped considerably. That is, only when speakers had very low language ability did they show brain waves that deviated from those of native speakers. Hence, proficiency plays a role for semantic processing.

Similar findings were also reported by Wartenburger and colleagues. In their experiment, participants were shown sentences that made no sense semantically (The mouse chases the cat) and compared them to sentences that did (The cat chases the mouse). The results revealed differences between high and low proficient late bilinguals. Interestingly, both groups showed increased activity in areas involved in cognitive control. The main distinguishing factor was that the high proficiency late bilinguals showed relatively more activity in the right fusiform gyrus, which as noted last chapter is involved in the processing of visual information (Devlin, Jamison, Gonnerman, & Matthews, 2006; Price & Devlin, 2003, 2011). Furthermore, lower proficiency bilinguals showed increased activity in the superior part of Broca's area. This area is known to be involved in the motor planning of verbal responses. Interestingly, low proficiency participants showed increased activity when processing meaning in their second weaker language. Thus those with higher proficiency are paying attention to the form of the words they are reading. Those with lower proficiency are busy trying to tie the words to their corresponding motor codes associated with each word (Wartenburger et al., 2003).

Taken together, the literatures on bilingual language processing and its brain bases suggest that for late learners there is an initial reliance on the first language. Over time, however, the second language begins to form direct links to concepts. However, this increased proficiency in the second language in adult learners does not reach that of monolinguals. Finally, the fact that proficiency influences semantic processing is consistent with the view that a late second language is learned in a more cognitive and academic manner, at least when this occurs in the classroom.

FROM EVERYDAY TO ACADEMIC LANGUAGE

So far we have reviewed evidence that proficiency involves an academic language component, what Cummins termed CALP. Because CALP involves the use of academic language, it is more likely to appear in tasks that require access to meaning and less likely to appear in tasks that involve grammatical processing. Furthermore, there is evidence that processing of grammatical violations is more sensitive to AoA in a second language. In the Cummins framework, we would consider grammar to be part of BICS and meaning or semantics to be a product of CALP.

In some ways, the notion that grammar is a part of everyday (BICS) language fits in with nativist theories of language development (Pinker, 1994). In this view, a three-year-old is seen as a linguistic genius. He or she has knowledge of grammatical rules without any instruction or guidance from parents or teachers. These rules are learned without the explicit instruction that plays a role in the academic learning environment of elementary, middle, and high school, where explicit teaching of language takes place. However, as we'll soon see, thinking of grammar as being influenced solely by BICS is not that simple.

Her interest piqued by BICS and CALP as explanatory measures, Lynette Austin chose a series of language tests that would help to quantify the language abilities of a group of Spanish-English bilinguals from the University of Houston. These tests included measures of academic language such as picture vocabulary and verbal analogies. She also included tests of grammatical processing in both languages. This included regular (jump, jumped) and irregular verbs (seek, sought) in English. Spanish affords a much more complex verb system. So rather than focus on regularity, Lynette chose complex verb tenses and sentence structures such as the conditional (If it happened...) and the imperfect subjunctive (If it were to have happened...) as well as simple verb tenses and structures like the present indicative (It happens). She also collected accent ratings and looked at the participant's ability to distinguish pairs of words in each language that differed by one sound (such as "pan, pun"). This gave Lynette a broad range of verbs that were both simple and complex.

The results revealed an interesting dissociation between AoA and language proficiency. Whereas AoA predicted accent in English (the second language) and the production of the regular past tense form, things that are learned early in life, proficiency was strongly related to the production of the irregular past tense in English. Spanish language proficiency was related to the production of advanced verbs in Spanish. Of particular interest was the lack of relationship between AoA and English academic language. These results indicate that AoA is a good predictor of things that are learned early in life such as the regular past tense and the acquisition of speech sounds.

Academic proficiency involves a more complex set of factors. First, academic language proficiency shows a moderately strong relationship across both languages. This is consistent with the notion of a single conceptual space as modeled in the revised hierarchical model of Kroll and colleagues (Kroll et al., 2010). Hence, the building up of complex forms of language benefits from general language knowledge in either language. These findings also shed light on our previous findings in the adoption literature. Although adoptees seem to lose access to word meanings and sentences in their first language, they are able to learn a second language with a high degree of proficiency. This is most likely due to the building up of CALP during childhood which then transfers over to the newly learned language.

ACADEMIC AND EVERYDAY GRAMMAR

Whereas semantics seems to have a direct connection to academic proficiency, grammar revealed a more complex picture. This can be seen in the dissociation between regular and irregular past tense in English. The regular past tense is associated with measures of everyday language. The irregular past tense was associated with proficiency as indexed by CALP. This suggests that grammar has components that are based on early language learning, which is the basis of everyday language. In addition, there are components of language that are built up over time. The results of Lynette Austin's dissertation suggest that grammar also has a complex academic component.

In chapter 4 we discussed evidence of proficiency as a modulator of the P600 in monolinguals. This wave is thought to index a higher level form of language processing that is important for sentence repair and for alternative interpretations. More interestingly, it appears to be a key determinant in the learning of grammar in late learners. Finally, this component is strongly related to language proficiency in a second language. Thus the strength of the P600 seems to be a reliable indicator of how proficient someone is in a language. The question that remains is how the transition to higher levels of grammatical ability occurs in second language learners.

Seminal work by Lee Osterhout and colleagues has begun to uncover the transition into higher levels of proficiency with classroom second language learners. In a series of EEG studies, second language learners were asked to look at sentences that had a grammatical error between the subject and the verb (I lives in Houston) or not (I live in Houston). These types of violations elicit a P600 in native speakers. However, in novice learners of a second language, an error of this type results in an N400. As we noted earlier, the N400 occurs in sentences with violations of meaning. Osterhout and colleagues suggest that initially learners of a second language use a word-based strategy. They think of *I live* as one thing. When *I lives* appears, an adult in the early learning stages compares it to the correct form *I live*. Since the two forms don't match, an early learner will see this as an error. As proficiency improves, learners begin to realize that there is an underlying rule-like behavior and hence no longer analyze these chunks as words. Rather they begin to break them up into pieces in a more analytical manner. For a more advanced learner, *I lives* is incorrect because the *s* should not be there.

The transition from a more word-based response to a grammatical response happens as speakers of a second language become more proficient across time. Interestingly, it turns out that there is a connection between standardized proficiency measures and the type of brain response observed. Those who have less proficiency in a language show stronger N400 responses. Those who have stronger proficiency in a language have stronger P600 responses. These results show us that late learners seem to apply an academic-like approach to learning a second language. Basically, they put word combinations into memory at first. Then they transition to decomposing them into a more rule-like combination. The learning of these rule-like combinations is what makes *I live* ungrammatical to fluent speakers of a language.

LANGUAGE AND MEMORY SYSTEMS

The notion of everyday language and academic language also aligns with the existence of different memory systems. Work in the memory literature has considered the nature of declarative and procedural memory. As I noted in chapter 4, declarative memory relates to things that we can talk about, such as past events or facts. Procedural memory is related to things that we do, both in terms of motor skills such as playing a sport or a musical instrument and in more cognitive domains such as chess, reading, and mathematics.

The possibility that procedural and declarative memory may help to explain how language is split up has also been considered in the second language literature. Paradis (M. Paradis, 2009) has offered a strong defense for the view that speech sounds, the sound combinations, and grammar are accomplished using implicit memory that is not subject to conscious awareness and is mostly procedural in nature. The processing of words and meaning is subject to conscious awareness and hence relies on explicit declarative memory. Paradis suggests a strong division between these two forms of processing in that a language function relies either on implicit memory or on explicit memory. Finally, the use of these different memory systems is regulated by maturation. During early life there is consistently more use of procedural memory. The appearance of declarative memory later in life comes to dominate the ways in which adults learn.

Paradis's model forms an interesting parallel with BICS and CALP. Everyday communication relies on a number of qualities that, at least on the surface, are procedural. They are not based on building up knowledge but rather on communicating in informal situations. Implicit parts of language such as accent are learned early in life. The preservation of native speech sounds in adoptees is another example of implicit procedural knowledge carrying priority early in life. The use of vocabulary, on the other hand, requires more explicit knowledge and hence is more likely to rely on CALP. According to Paradis, this type of knowledge develops later in childhood, when declarative memory comes to dominate learning.

Using the same basic constructs as Paradis, Ullman has proposed a neural model that takes into account procedural and declarative knowledge. Procedural knowledge involves brain systems in the basal ganglia (which includes the thalamus, putamen, and caudate nucleus) and in areas in the frontal cortex involved in motor planning such as the premotor cortex and Broca's area. The declarative system involves the hippocampus in the medial temporal lobe as well as areas in the frontal cortex. H.M., whom we talked about last chapter, had damage to the medial temporal lobe and lost the ability to form new conscious memories. Areas just below and in front of Broca's area are involved in the processing of meaning. Ullman suggests that these systems are involved in first language processing, where the processing of words and their meanings are handled by declarative memory, and procedural memory is needed to combine and recombine pieces of words in order to form grammar.

Both procedural and declarative memory also apply to late second language learning. During the initial stages of second language learning, the declarative memory system plays a stronger role. However, over time as proficiency in that language improves the ability to process information becomes proceduralized. Hence, second language learners should begin to show similar underlying signatures in the brain compared to first language learners. This view fits in nicely with the research we discussed earlier by Osterhout and colleagues. Namely, second language learners begin by using a meaning-based strategy during initial stages of language learning. However, over time they transition to using more of a native-like processing system as indexed by the P600, the positive going wave that appears when grammatical processing becomes difficult.

The declarative/procedural distinction lines up in an interesting way with CALP and BICS. Everyday language occurs in a more implicit procedural way. There is little intention to learn things, and they just occur naturally. CALP requires more explicit instruction and thus comes to rely on declarative memory structures. Ullman's model suggests that these two forms of processing may rely on different neural systems. However, to date little research has linked the concepts of BICS and CALP to the type of memory in a more direct manner.

SPEECH IN A SECOND LANGUAGE

As I noted in chapters 2 and 3, speech is clearly linked to early development. In addition it is part of the procedural learning system in that children learn it without much conscious intention. Finally, speech is used in everyday life both in informal and formal contexts. Thus speech at the basic level is part of BICS since it is used every day for communication.

Many bilinguals however do not learn their dominant language at birth. Many are exposed to a language later in childhood, adolescence, or in adulthood. In this group of individuals, speech could take advantage of the procedural system or it could involve declarative memory systems as well. This was the main question that Pilar Archila, a graduate student working in my lab in collaboration with Jason Zevin of Cornell Weil Medical School and Ferenc Bunta from the University of Houston Speech and Communication Disorders department, set out to ask in a series of studies conducted in my laboratory.

In these studies, Archila sought to look at the influence of AoA and proficiency on the processing of speech sounds (Archila-Suerte, Zevin, Bunta & Hernandez, 2012) The study, which I discussed in chapter 3, presented participants pairs of sounds that shared ("saf"-"saf") or differed in a vowel sound ("saf"-"sef"). Archila asked a group of early and late bilinguals to listen to pairs of sounds that matched or did not match. AoA played a role in how tightly clustered a set of syllables would be. For monolinguals and early learners with high proficiency, there were four distinct clusters that could be associated with each sound of the four syllables presented, "saf," "sef," "sof," and "suf." However, for late learners and early learners with low proficiency the four clusters that represented each sound where much more spread out. Hence, proficiency in a language was not associated with tightness of the clusters.

In order to quantify the distance between each cluster, Zevin developed a second metric that quantified how a participants decisions differed with regard to the distance between the center of each cluster (referred to as between-category variance). Archila-Suerte, Bunta, Zevin, and I interpret this as relating to a more cognitive form of speech recognition that relies on hearing the difference between different speech sounds. In this view, late learners do not have speech sound categories formed the way that early learners and monolinguals do. Rather they arrive at second language learning later in life and must adapt their first language speech system to fit that of the second language.

One way to conceptualize this is to think of an analogous domain, perception of musical notes. The ability to detect a single note in isolation, called absolute or perfect pitch, is present in a small group of people. In general, the ability to learn this is highly dependent on age. Work with musicians has shown that absolute pitch can only be learned by speakers of nontonal languages before the age of seven (Deutsch et al., 2006; Trainor, 2005). However, perfect pitch is not a prerequisite for musical success. Many individuals learn to recognize a note by comparing it to another note. The ability to recognize via a comparison process is called relative pitch.

The recognition of notes in musicians that use relative pitch is similar to the recognition of speech sounds in second language learners. Rather than being able to recognize single sounds in isolation, late learners of a second language rely on hearing the differences between sounds. Most importantly, the distance between these perceptual-decision clusters is correlated with language proficiency. In short, it appears that late learners have a different way of processing sounds that involves the use of a higher level cognitive strategy. Hence, comprehending speech in highly proficient late second language learners can be thought of as involving a more declarative and cognitive approach.

We were also interested in looking at how the brain of bilinguals adapts to speech perception in the second language. To do this, Archila asked groups of bilingual children and adults to listen to a speech sounds while watching a movie with the audio soundtrack muted. Using functional MRI, which measures the brain's vascular response, she tested a group of children who learned English during kindergarten but were between the ages of 6 and 10 at the time of testing. She also tested a group of monolinguals of the same age as well as monolingual and bilingual adults with similar learning histories.

The results revealed findings that were in line with what has been observed in the literature. First, monolinguals showed the expected pattern of results. There was increased activity in the bilateral superior temporal gyrus, in those areas involved in the processing of speech, in both adults and child monolinguals. Most importantly, there was no significant difference across age groups. This suggests that the basic speech recognition mechanism is in place very early in life and does not change substantially across age when a person is only exposed to one language.

The pattern of results showed interesting similarities and differences in the bilingual group. The youngest group of bilinguals showed increased activity in the superior temporal gyrus, a pattern similar to that seen in monolingual children. However, by age eight, bilinguals began to differ from both the younger bilingual group and monolinguals. In this case, there was increased activity in speech areas but also in the prefrontal cortex and the inferior parietal areas of the brain both on the left and right. This pattern changed again in adulthood. Bilingual adults with a similar learning history did not show this increased activity in nonspeech areas. Hence, recruitment of areas in the frontal and parietal cortex was most likely due to the need for additional cognitive resources when learning the speech sounds of the second language. Taken together these results suggest that learning a second language during childhood results in an adjustment process. Specifically, the brain of children must recruit areas involved in working memory and cognitive control to understand speech in the second language. As time passes and bilinguals transition to higher proficiency in adulthood, these differences largely disappear and bilinguals come to resemble monolinguals in their activity to speech sounds in English.

Archila was also interested in looking at how adults differed in their brain activity across high and low proficiency groups. As noted earlier, there were clear differences in early and late learners in a variety of areas involved in timing and auditory speech processing. These areas were restricted to the sensory regions of the brain, with early learners showing much smaller areas of activity relative to late learners. High proficiency was associated with increased activity in speech areas of the superior temporal gyrus and audiovisual integration areas in the precuneus area of the parietal lobe. Low proficiency, on the other hand, was involved in a wide swath of multiple brain areas. This included separate portions of the middle temporal gyrus on both sides of the brain, which indicates more complex processing of auditory speech. Interestingly, there was also considerable activity in the frontal lobe as well. Hence, academic proficiency helps to sharpen up the brain areas involved in language processing independent of whether a language is learned early in childhood or well into adulthood. The interesting part is that in second language learners there is a relationship between academic proficiency and speech. This suggests that speech in the second language, like grammar, can come to rely on declarative memory.

PITRES'S RULE

We began this chapter by considering the case of a 22-year-old woman who had learned English at the age of 10. She was operated on in order to remove a defect in her brain in the temporoparietal area. After her surgery she had a stronger impairment in Spanish, her first language, than in English her second language. Katheryn Kohnert felt that the Boston Naming Test, which is the only test that was administered in both languages, might not have adequately captured language proficiency in both languages. First, this test did not include items that may not have the same frequency in both languages. For example, items such as "trellis" and "sled" are not readily talked about in most Spanish-speaking countries. Hence, scores even in monolingual Spanish speakers might be depressed relative to monolingual English speakers. This problem exacerbates the one mentioned above. Basically, there are many reasons besides brain damage that the patient might have scored 32 out of 60 on the second try. Having only half the data for Spanish before the damage just makes interpretation that much harder. Most importantly, there was no comparison group of normal bilinguals that could be used to determine whether the pattern of results observed was indeed abnormal.

In order to gain a better idea of what an unimpaired group of bilinguals around the age of 22 would score on the Boston Naming Test, Kohnert decided to collect her own data with this group. She tested a group of 100 sequential Spanish-English bilinguals who learned English in middle childhood. The data revealed a pattern of English dominance in this group. The mean score in Spanish and English was 32 and 47 out of 60, respectively. In addition, Kohnert calculated the 95 percent confidence intervals that let her know where 95 percent of the scores should lie. These intervals ranged from 15 to 49 in Spanish and 34 to 59 in English. In short, we would expect a normal bilingual without brain damage to be English dominant on average. Most importantly, truly deviant scores would be below 15 or above 49 in Spanish and below 34 or above 59 in English. These results suggest that the patient tested by Gomez-Tortosa (32 in Spanish and 44 in English) was not outside of the range observed in normal Spanish-English bilinguals who learn English in school. The fatal flaw in their study was not taking into account the nature of proficiency in the underlying normal population. Hence, proficiency plays an important role when considering the effects of brain damage on language processing in bilinguals. It is important to note that the AoA of our participants was slightly lower than that of Gomez-Tortosa's patient. Nevertheless, our results show just how much individuals can vary in their language proficiency profile. I suspect that the patient identified by Gomez-Tortosa was showing preservation of the more familiar language. This pattern of breakdown observed by Gomez-Tortosa and colleagues fits Pitres's rule, which suggests that the more familiar language may be more resistant to breakdown after brain damage.

LANGUAGE LEARNING AS A BUILDING PROCESS

This chapter has also discussed examples of changing proficiency across early childhood. This suggests that the language environment can influence how well a person can speak a language. It can also provide a seed for a language that can later be recovered even in adulthood. The disjointed nature of language acquisition is something I have experienced on a personal level. During my initial learning of Portuguese at age 20, I had a relatively strong Spanish accent. This accent was readily apparent when I arrived in Brazil, as I constantly was asked if I was from Argentina. Over the course of months I watched as my thick Spanish accent peeled away. The verbs and variation of word order from Spanish came in handy. The *z*'s and *zh*'s of English helped to smooth my accent. The recovery of my childhood memories for Brazilian music didn't come right away. But once it took root it was enough to make me sound like a native. Even many years later, people wonder how I learned Portuguese. From this perspective, you can clearly see how learning is building.

In this chapter, we have considered the nature of proficiency as it relates to learning two languages. What emerges from this literature is a clear impact of proficiency across a wide swath of different language processes. This includes semantics, grammar, and even speech processing. In short, proficiency touches every aspect of language processing. The interesting question to consider here is how language learning in bilinguals informs us about expertise in general.

One way in which the learning of two languages expands our thinking about expertise is that it enhances the nonlinear aspect of this process. As I noted earlier, deliberate practice as conceptualized by K. Anders Ericsson has had a profound impact on the way in which people view expertise. Number of hours is an excellent proxy that helps to capture the importance of continued development across time. The problem is that when it comes to language we can see how this process is much more complex than that. Early sensorimotor learning that is necessary for learning speech sounds shows some resilience across the life span. Early exposure to language can result in learning that may be lost but can later be recovered. Furthermore, learning in one language seems to readily transfer to a second language. This suggests that most language learners do not start from zero.

One thing that should also become apparent is that both Pitres and Ribot were right. The age at which something is learned is as important as the degree of familiarity. If learning of a second language happens at age five, then the brain takes whatever resources are available and applies them to learning a second language. But this learning also depends on how much time is spent on the process. Over time deliberate practice will continue to build up expertise in one or two languages. However, the brain is incredibly resilient, and much can be relearned even at points during which it was thought a memory or a skill had been lost entirely. If we apply the notion of 10,000 hours and push it further across time, it suggests that a person could indeed relearn a language later in life to very high degrees of proficiency. In short, the work in this chapter suggests that language learning is a long, drawn-out process that requires many hours to firmly take root. Familiarity comes in many guises with potential curves and turns. But in the end proficiency, whether it be from an everyday component or from a more academic approach, comes to play a crucial role in the brain systems involved in processing a second language.

NOTES

1. These sentences were taken from a Linebarger, Schwartz, and Saffran study originally conducted with adults suffering from a language impairment.

2. This group presents an interesting wrinkle to the revised hierarchical model in that participants become dominant in a second language. These effects become reversed, suggesting that the most proficient language is not always the first language .

6

CONTROLLED ADJUSTMENTS

IN THE FIRST two sections of the book I traced out how learning is influenced by two particular factors: age of acquisition (AoA) and proficiency. I also traced out how these two factors work together to constrain learning in general and language learning in particular. These extensions of Pitres's and Ribot's rules are important and fundamental to the bilingual brain. However, consideration of these two factors in isolation falls short. Work with bilingual aphasics has confirmed the importance of a third factor that was initially identified by Pötzl. The nature of cognitive control (also termed "executive function" and/or "working memory" by other researchers) is reflected in the initial cases he reported, in which a bilingual appeared to be stuck in one language and unable to shift out of it. In this view, the language was unreachable because of an inability to switch to the appropriate language. Hence, the problem was not with the loss of language. Rather the issue was an inability to access a language.

Just as proficiency and AoA have parallels outside of bilingualism, cognitive control and switching also show similar parallels. Examples of switching and control abound in the real world. The most common example involves all of the multiple tasks that people engage in when driving. It is now well documented that engaging in any kind of alternative task while driving results in an increased cognitive load and distraction (Drews, Yazdani, Godfrey, Cooper, & Strayer, 2009; Strayer & Drews, 2004; Strayer & Johnston, 2001). However, many other daily activities also require control and planning. I have noticed it when cooking. I used to get the order of things all wrong. I would start working on the main entrée (maybe some fish or chicken) and then turn to do the side dishes. This makes sense from a cognitive point of view because we think of entrées as being the most important thing. The problem is that the entrée does not always take the longest to cook. So what ended up happening at times is that I would have the entrée done, but the rice would not be ready yet. To avoid this I began to spend a little time at the beginning thinking about how long it would take to cook each part of the meal. I could then sequence things so that the whole meal was done about the same time. I quickly realized that most of the time, I had to start the rice or potatoes first, followed by an entrée, and then by the vegetables.

Cooking is probably a hybrid activity in that it may involve language but it may not. But there are many instances in which people have to switch between tasks that do not involve language. For example, when you have to react to a car that is changing lanes in front of you while observing traffic behind you in the rearview mirror. This form of switching may involve little language. Hence, it is plausible that the ability to switch between languages may rely on neural and cognitive mechanisms that were designed to handle these types of tasks in a single language or even without language. In this chapter, I'll focus on the ways in which cognitive control emerges outside of a bilingual context. I will also trace out how control develops during childhood and into adolescence both in terms of the brain mechanisms that are involved in this process and the concomitant behaviors that are observed. This will set the stage for the issues of cognitive control within a bilingual context that I'll discuss next chapter.

CONTROL AND THE BRAIN

One of the hallmarks of human cognition is our ability to control our actions (for good or bad). Let's take the example of dogs. I don't have the best experience with dogs. I have had them jump on me at beaches. One time I was driving home and two dogs came out from behind a fence and proceeded to chase my car down the block. I didn't want them to follow me all the way home. So I quickly drove through some side streets and lost them before going back home. To top it off, my neighbor's dog continues to bark at me after eight long years of taking out the garbage or opening the garage. By now you'd think he would understand that I am not going to respond. At times it still annoys me. All sorts of scenarios go through my head about how to silence him (most of them unmentionable in these pages), but before I get to the silence I can quickly see how each of these leads to a less than ideal outcome. The interesting part of this is that my neighboring canine would never be able to navigate the complex decision making and control necessary to hold his bark. So I continue using mine to imagine silence and hope that one day it might actually happen.

Many of these differences between humans and other animals are most likely attributed to the size of the prefrontal cortex, the most anterior portion of the frontal lobe.

As we move away from the central sulcus, our brain becomes less and less tied to simple motor processing. The first set of areas directly in front of the motor strip have been termed as premotor areas. These areas are involved in the planning of movement. In language processing, Broca's area is a motor planning area that is involved in speech.

As we move further forward we encounter the part of the brain that begins to distinguish us even further from the rest of the animal kingdom. Here we encounter the prefrontal cortex. In fact, a quarter of our brain is prefrontal cortex. Compare this to 15 percent in chimpanzees, 7 percent in dogs, and 4 percent in cats. So at least some of the difference between my neighbor's dog and me is most likely due in part to the respective size of our prefrontal cortex.

The importance of the prefrontal cortex can be seen in the case of animals who have been domesticated to act like humans. One of the most famous cases is that of Nim Chimpsky. Two linguists who thought the environment played a very strong role in determining the differences between chimpanzees and humans decided to raise both together (Hess, 2008; Terrace, 1987). Nim Chimpsky (a play on the name of the famous linguist Noam Chomsky) was placed in a human family by a research team intent on showing that the only difference between chimps and humans was environmental. They thought that a chimpanzee would become human if simply allowed to be raised like a human child.

The experiment failed miserably. Nim never really acquired the richness of language that is seen in humans. He could produce single words but was unable to weave them together using the complex grammar that a three-year-old produces on a daily basis. It was clear that he understood humans better than any other chimp. But he was never able to navigate the complex social rules of modern human society.

So what might have held him back? As I intimated earlier, the answer lies in the brain and specifically in the size of the prefrontal cortex. Researchers who take an evolutionary approach have looked at higher primates for clues about how brain expansion may have influenced how humans interact with the world. Two areas that are larger in humans relative to other higher primates are the ventromedial portion of the frontal cortex and the dorsolateral portion of the frontal cortex. The ventromedial cortex is connected to inferior posterior temporal regions of the brain involved in object perception and the limbic structures that are involved in the regulation of internal states. The dorsolateral prefrontal cortex, on the other hand, is connected to regions in the parietal cortex that are involved in multisensory integration and the associated areas in the dorsal visual stream that are involved in the location of external objects. Hence, the increase in the prefrontal cortex involves not only increased real estate but also an increase in the richness of subareas that are highly interconnected with other regions of the brain. These interconnections allow humans considerably more control over lower cortical centers relative to other animals.

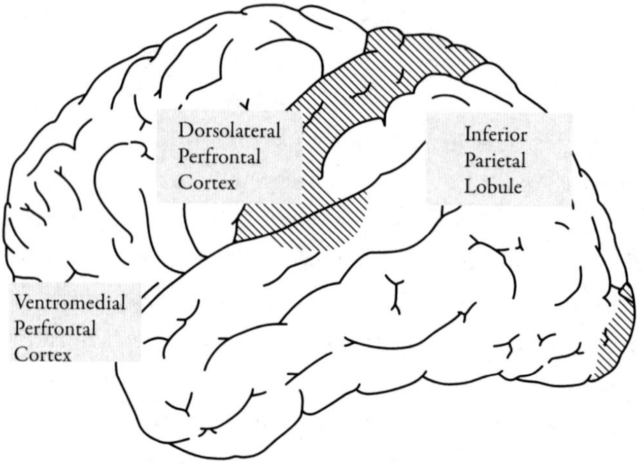

FIGURE 6.1 Rendering of the cortical brain areas involved in cognitive control.

Particularly interesting are the cognitive styles associated with these two processing streams. This can be seen in the ways that differential neural development of each area is associated with different cognitive styles of orangutans, bonobos, and chimpanzees. Orangutans, who have a relatively solitary lifestyle and less sophisticated social organizations, have relatively less developed ventromedial prefrontal cortices. The importance of the ventromedial area in social interaction is also seen in the difficulty that human patients with lesions in this region have interacting in socially appropriate ways (Bechara, Tranel, & Damasio, 2000). This region has also been found to play an important role in evaluating our own internal state and using that to form inferences about what others are thinking and feeling (Stout, 2010). Taken together the results from a cross-species comparative approach suggest that the ventromedial prefrontal cortex is crucial for social reasoning.

The dorsolateral prefrontal region, like the ventromedial prefrontal cortex, also shows differential degrees of development across species. It is larger in orangutans and chimpanzees than it is in bonobos. Interestingly, the bonobo is the only species of the three that does not construct tools (van Schaik, Deaner, & Merrill, 1999). Together with other data gathered from human imaging and neuropsychological studies, this suggests that the dorsolateral region is important for the control of external actions. It is also important for instrumental action and the need to have control in order to get distinct parts of a motor routine and its effect on the environment worked out.

The impressive part of human cognition is that all these systems are larger and more developed relative to the higher apes. This not only enables us to have very sophisticated social structures but also allows us to construct sophisticated ways to interact with a world full of artifacts and objects. Our ability to navigate a very complex environment is one of the reasons that Nim Chimpsky was unable to become human. With an upper limit on the real estate in the prefrontal cortex, he was at a permanent disadvantage for his entire life. The difference between a chimpanzee and a human is not due entirely to differences in how they are raised. It also has to do with the profound differences in the size of the prefrontal cortex in each species.

HOW DO WE LEARN CONTROL?

The importance of the prefrontal cortex and its influence on control is also mirrored in the stages of cognitive development. The father of modern developmental psychology, Jean Piaget, had a very clear view about how cognitive skills including language developed in humans (Piaget 1970). In his view, humans were born with a few basic reflexes.[1] The first few months were relegated to interacting with a purely sensorimotor world filled with images, sounds, and a few basic ways to engage with them. In fact, some researchers argue that the infant world is so undifferentiated that it does not even honor the distinction between senses. For example, infants will get bored looking at a light that flashes at the same rhythm as a tone played over speakers (Maurer, Mondloch, Robertson, & Sagiv, 2005). Adults, however, do not show such an effect. One interpretation is that infants actually confuse information across their senses much more readily than adults. This confusion, called synesthesia in adults, is often seen in those with artistic inclinations and with some savants (Maurer et al., 2005). However, in the majority of individuals there is no confusion between what a person hears and sees. Over time, normally developing children learn to differentiate these senses.

During the first few months, an infant is almost stuck in the world of the here and now. The focal nature of his or her existence is limited in time and space. He or she can only react to what is in front of them at that very moment. Hence, these interactions are relatively short-lived. However, within a few months this bubble bursts; and soon infants know that out of sight does not mean out of mind. Over time a child begins to develop more and more sophisticated ways of pairing actions with consequences.

According to Piaget, the ability to know something exists without seeing it is one of the hallmarks of development. Piaget first established this by asking infants to search for an object that was in plain sight (Piaget, 1954). Then he would cover the object to see if infants would still search for it. What he found was that until about eight months of age, infants would not search for an object that was covered up. It was as if the object was out of sight and out of mind. Around 10 months of age, children began to search for an object even when it was hidden. Subsequent research has suggested that object permanence appears much earlier in infancy (Baillargeon, 2004). Despite the debate about when infants are able to represent objects in the world with their minds, researchers agree that changes in early life involve an improving ability to keep objects in mind even when they are not present in the world.

In Piaget's theory, object permanence played an important role because it was a reflection of the fact that toddlers were able to "represent" the world in their minds. I experienced this firsthand with my oldest daughter, Kiara, when I worked at home every Friday during her first year of life. For the first six months or so I could play with her for a few minutes and then walk away. When I left, she would continue happily playing with her babysitter as if I had never been there. A few months later I played with Kiara for about 15 minutes then walked away as I had done many times before. This time she followed me with her eyes and called for me in the way a wordless 10-month-old does. I had gone from fleeting image to a permanent dad in the span of a few months. If we fast forward 10 years, my youngest daughter, Kamille, at 15 months would often walk into our room to look for her mother and me, even when we were not at home. She knew that we existed even if she could not see us.

At the time that Piaget carried out his research, he did not couch his framework for cognitive development within a theory of brain development. However, more recent research has begun to consider the link between object permanence and the neural changes occurring during early life. Yuko Munakata is one of a handful of researchers who have started to rethink what Piaget's stages mean in terms of brain development (Shinskey & Munakata, 2003). Along with Jane Shinskey, she sought to uncover the root of this effect by finding a way to see inside infants' minds. They started by replicating an experiment that dates back to the days of Piaget. To do this they placed 6.5-month-old infants in front of toys that were covered by a cloth or not. They found that infants were more likely to reach for a toy if it was in plain sight than if it was covered by a cloth. Next they placed a toy in front of the child and then gradually dimmed the room to complete darkness. What they found was that infants would reach for an object that was covered in darkness to a greater extent than when the object was covered by a blanket in plain sight. What these results show is that the difficulty with finding a hidden object was due at least in part to the fact that the infant was distracted by the real world. Hence, the ability to see an object that is covered depends on keeping two things in mind, what they see and what they don't see. When the lights are off, infants cannot see anything at all. So their performance actually improved because the mind can create an image without the real world getting in the way.

The importance of keeping two things in mind is crucial for our everyday existence. Let's take a crude adult analogy to illustrate this point. Let's say you are getting ready for a very important appointment and things are not going as smoothly as you would like. Maybe your coat is not quite falling into place or your pants are showing that crease you tried desperately to iron out a few minutes ago. As you are about to leave you realize that you left the file with a bunch of important papers in your home office. You walk over to your home office with the full intention of getting your papers. But just as you arrive the phone rings and you pick it up. It is a recording reminding you of your yearly checkup tomorrow at 3 p.m. You quickly jot it down in your agenda and head to your appointment. As you are ready to drive off you realize that you once again forgot the file.

Why did you forget? Just like our infant above, you got distracted. One goal entered your mind (noting the checkup) and overrode the second goal (getting the file). The difficulty arose because of the burden of having to switch between the thing you are doing in the world right now and what you need to do in the world later. Shinskey and Munakata's results suggest that infants face a similar type of situation when involved in object permanence tasks. They are searching for an object that is present in the world but hidden from view. In order to find the hidden object, they need to ignore the world and attend to what is in their minds.

The results of research show that infants initially relegated to a sensorimotor world of fleeting images and simple movement during the first few months of life quickly transition to a concrete world by the end of the first year. By the end of the second year children moved past this sensorimotor stage and emerge with the ability to use and understand symbols. Even before speaking, infants can associate hand gestures with the real world. It is well known that most 10-month-olds can learn simple signs to communicate.

The most profound use of symbols in humans appears around a child's first birthday when they begin to use language. Although object permanence and language may not seem to have much in common, according to Piaget there is a very strong connection. Specifically, both of these activities involve placing an object that is not in plain view in one's mind. By using a single word, a child can instantly bring an idea, concept, or object into his or her own mind and into someone else's mind. Symbols are a way of describing the world without directly referencing its physical characteristics. Language allows us to communicate without playing charades.

The symbolic nature of language can also be seen in its arbitrariness. Although the word "buzz" may sound like the noise a bee makes, the word "bee" has no clear relationship to the object bee. But children begin to learn all sorts of arbitrary connections between the sound system they have developed and the world that adults are talking about. These connections require children to go beyond the world that exists in front of them.

From the first birthday on, normal children go through a dizzying array of changes. These changes spread beyond language as a child begins to develop a sense of self. Liz Bates, my doctoral advisor, did her early work with children, looking at how symbols emerge at this young age (Bates, Benigni, Bretherton, Camaioni, & Volterra, 1979). She realized that children were very good at representing the world in their third year of life. For example, she would ask them to play a game in which they would use the conditional. Children could imagine all sorts of things. They were happy to imagine that the moon was made out of chocolate. They could even imagine that they were an animal or someone else. Then came the clincher. She would say to them, "Imagine that you were you." This left some kids in tears. "Of course I am me. Who else would I be?" So symbolic representation only goes so far, at least at age three or so (Bates, 1974).

Liz Bates's early work points to another remarkable milestone in human development. Children begin to understand that their view is different from that of others. Piaget termed the inability to do this "egocentrism." He thought it was impossible for young children to understand that what they saw in the world was different from what another person saw. Piaget tested this using a three-mountain problem, in which an experimenter and a child would sit on opposite sides of a display. The experimenter would then ask the child which display matched his or her view. Inevitably, the child at first would choose his or her own view as the one that matched the experimenter's view even up to the age of five. By age seven, most children would begin to choose the experimenter's view. This general trend is observed in another series of tests, in which children are asked to reason about the physical world. For example, young children who are shown a cat with a dog mask will say it is a dog. Kamille, my youngest daughter, would stay very close to us when we used to visit the Halloween store. We insisted that the mummies, skeletons, and ghosts were just big motorized mannequins. She refused to even touch or look at them. A couple of years later she magically understood. "They are just pretend," she would say. Like Kamille, children around school age begin to understand the difference between appearance and reality.

Researchers extended Piaget's original finding to consider whether children can reason about what other people see compared with what they see. In these false-belief tasks, children are asked to observe a scene (Wimmer & Perner, 1983). In the first scene, a toy is placed in one drawer while a small puppet named Tommy is viewing it. The experimenter then lets the child know that "little Tommy" has left the room. The experimenter then moves the toy to a different place in the room such as under the bed. When little Tommy comes back the child is asked to indicate where he will look for the toy. The key to this task is that the child has to understand that little Tommy thinks the toy is in the drawer not under the bed. However, young children will indicate that Tommy will look under the bed. That is, they are unable to distinguish between what someone else knows and what they know. Only by age four or so will they begin to realize that Tommy will look in the drawer (Wimmer & Perner, 1983). In short, only by age four are children able to realize that others think differently than they do.

As I noted earlier, many of these tasks involve cognitive control, in that they rely on children keeping one thing in mind while processing another. Much like me in the example I used above that involved having to keep the files in mind in the face of distraction, children find it difficult to focus on something that is counter to what is in their mind. In both these cases, cognitive control is important in order to allow an individual to focus on the right piece of information. In short, cognitive control is crucial for development.

The ability to imagine things that are not real continues into adolescence. As puberty approaches, humans begin to understand that others see the world differently than they do. Furthermore, they begin to create thoughts that are more and more divorced from physical reality. Abstract thought is a reflection of a child's ability to think beyond the physical world that exists. In fact, at the end of childhood the advent of abstract thought appears as discussions in classrooms turn to topics such as truth, honesty, or liberty. These concepts are not directly based on the physical attributes of the world.

Based on all his work with children, Piaget suggested that across development humans become less and less bound by the physical world. Specifically, adolescent thought is marked by a distinct ability to imagine what is not true. It is unlikely that an adolescent would end up in tears when asked what if they were themselves. Consider how often you have thought, "I wasn't myself today." The dissociation of ourselves is perhaps the greatest skill we possess and definitely relies on the expanded power of our prefrontal cortex (Uddin, Iacoboni, Lange, & Keenan, 2007).

BRAIN DEVELOPMENT AND CONTROL

These four stages of development also line up nicely with brain development. Areas of the brain involved in sensory and motor activities are the first to develop. Subsequent maturation adds to the function of these basic areas as a child grows older. As this happens, areas that integrate different sensory and motor systems begin to mature. The final cog in our cerebral wheel comes with the development of the prefrontal cortex, which lies directly in front of the basic motor areas of our brain. The prefrontal cortex, which is thought to be crucial for delayed gratification and higher level reasoning, does not develop fully until age 18. This late developing area is also thought to be important in reasoning about the real (concrete) and abstract (formal) world. The progression from sensorimotor to abstract thinking observed by Piaget shows an interesting parallel with brain development (Gogtay et al., 2004).

FROM CONTROL TO FLEXIBILITY

As I outlined earlier, work in the developmental literature has begun to form a link between executive function, which is mediated by the prefrontal cortex, and the ability to form internal representations. Interestingly, a parallel line of research has developed with regard to cognitive flexibility. What is cognitive flexibility? Think about the example I gave earlier about having to pay attention to multiple sources of information while driving. This requires flexibility. For example, a driver might notice that the car in front of her is speeding up and speed up to match her speed. At the same time, she may notice that the car coming by her is about to change lanes. A driver in a good mood will slow down a bit allowing the car to switch lanes. As she is switching between what is in front of her and what is behind her she might hear her favorite song on the radio and turn up the volume. The ability to switch between these different forms of information is what researchers have come to term "cognitive flexibility." A person with little flexibility would have trouble switching between driving and adjusting the radio. Hence, cognitive flexibility is the ability to use control to switch between different sources of information.

The nature of switching has also been the topic of extensive study in the adult literature. Like many discoveries in psychology, the presence of switching effects was first noted by an unlikely candidate. Seminal studies exploring the nature of task switching were conducted over 80 years ago in order to look at which conditions yielded a cost and which ones did not (Jersild, 1927). To do this Arthur Jersild presented stimuli in a pure condition that required only one task or in a mixed condition that involved having to perform a different task on every other trial. Interestingly, mixing tasks did not always lead to a slowing of responses. For example, when participants had to see a two-digit number and a word (i.e., 64 bad) and were instructed to subtract 6 from the number and produce the opposite word, they were actually faster than when just viewing words or numbers. Notice that in this task that each stimulus (i.e., the word or number) can only yield one operation. You cannot subtract 6 from the word "bad." However, when asked to alternate between adding 6 and subtracting 6 for separately presented two-digit numbers there is a substantially larger cost than when having to perform only one of these tasks. The key to this effect is that the two-digit number triggers both tasks (you can subtract or add). The participant has to choose one of these tasks and then switch on each subsequent trial.

Arthur T. Jersild went on to have a very prolific career in educational psychology. It was only very late in his career that a revival of these switching studies occurred. Spector and Biederman (1976) sought to use more modern experimental techniques to address some of the potential limitations of the original studies. For example, they presented each stimuli independently rather than using a list of items on a single sheet of paper. Since then a number of studies have replicated the costs of switching between tasks relative to performing a single task (Allport, Styles, & Hsieh, 1994; Hsieh & Allport, 1994; Jersild, 1927; Rogers & Monsell, 1995; Spector & Biederman, 1976). Although switching costs appear for both errors and response latencies, the majority of studies index costs in terms of latency.

The effect of task switching has an interesting implication in the increasing use of mobile phones while driving. As I noted earlier, a number of studies have found that driving while using a cell phone leads to a substantial inability to navigate adequately (Drews et al., 2009; Strayer & Drews, 2004; Strayer & Johnston, 2001). This has led many governments across the world to ban the use of hand-held units in favor of earphone or loudspeaker hands-free systems. However, subsequent studies have found that the use of hands-free units do not lead to any reduction in the cognitive load experienced while driving. Interestingly, there appears to be much less interference when speaking to someone in the passenger seat. Similarly, people hear music and sing along all the time with few problems. Personally, I felt that using the cellphone had a strange effect of splitting where I thought I was. It was almost as if I was in two places at once. What I wondered is whether this physical displacement might play a role in the distraction suffered by drivers on their cell phones.

In 2008 Fabio Ferlazzo and a group of his colleagues asked participants to use either a loudspeaker or an earphone piece while performing a series of visual tasks (Ferlazzo, Fagioli, Di Nocera, & Sdoia, 2008). Their notion was that the perceived distance of the voice would play a role in how well they could perform the visual attention task. Their results confirmed their suspicions. Participants were worse at tracking stimuli on a computer screen when they used an earphone system compared with having a conversation over the loudspeaker. In short, it is as if the use of earphones made the conversation proximal and intimate. This had the effect of crowding out things that were further away. However, conversations over a loudspeaker were perceived as being further away. This allowed a participant to keep both the computer screen and the voices away from them. Ferlazzo and colleagues suggest that using an earphone or holding a phone to one's ear has the same effect. The best solution is to use a voice over the loudspeaker, allowing a driver to experience both the road and the conversation as being further away. In short, keeping both things at a perceived distance improves the ability to do two things at once. Like the task switching literature, it appears that doing two things at once is not always impossible. So if you are going to perform two tasks at once you might want to make sure that the two tasks are incompatible (like simple numerical operations or opposites). Most importantly don't make your secondary task (i.e., talking on the phone) so close to you that it effectively crowds out the road in front of you.

CHANGES IN TASK SWITCHING ACROSS DEVELOPMENT

The notion of task switching and cognitive control shows an interesting parallel in both the developmental and the cognitive psychology literature. Earlier we discussed examples of how having to find a hidden toy or imagine that someone sees something different from what you see involves having to attend to an internal stimulus and ignore an external one. In a similar fashion, task switching involves having to attend to a certain aspect of one stimulus and ignore another. If one has to subtract the number 6 from 25 on one trial, then one would have to ignore that you added 6 on the previous trial. On the subsequent trial one would do the reverse.

In the 2000s, researchers began to use the same switching tasks that were revived in the 1970s and 1980s to look at changes across development. One of the most comprehensive studies in this respect was conducted by Davidson along with Adele Diamond (Davidson, Amso, Anderson, & Diamond, 2006), a researcher with a long-standing interest in the developmental changes that allow individuals to perform better and better on cognitive control and switching tasks. To explore this, the researchers gave participants three different versions of classic cognitive control tasks. These three tasks involved presenting either circles, pictures, or arrows. In each case, the stimuli would signal a response either on the left or right side of the screen. For example, in the dots test a filled dot would signal a left button response and a striped dot would signal a right button response. In addition, the filled or the striped dot can be placed on the same side as the response or on the opposite side of the response. That is, a filled dot on the right side of the computer screen would be incompatible, whereas the same dot on the left side is compatible. Results from many studies find that individuals slow down when the stimulus signals a button press on the opposite side of its physical location on the computer screen. Finally, Davidson and colleagues also presented the stimuli in a mixed format, which involved having items switch from congruent to incongruent and vice versa.

The results revealed considerable changes across age. First, younger children tended to show a larger incompatibility effect. They really slowed down and made many more errors when the dot signaled a response on the opposite side. Furthermore, slower responses and more errors were observed when children had to switch across a task (i.e., make compatible and incompatible decisions) versus when they did not have to switch (i.e., make only compatible or incompatible decisions in one series of stimuli). The cost of switching diminished as children got older and became smallest in adults in their 20s.

Given that task switching shows a reduction across development and that the frontal lobes are also known to show an increase in myelination from childhood to adulthood, it is natural to think that both of these facts might be directly related. To test this, Crone et al. (Crone, Donohue, Honomichl, Wendelken, & Bunge, 2006) asked groups of children, adolescents, and young adults to perform tasks that either switched or did not while being scanned with fMRI. Furthermore, they made the tasks have single mappings or double mappings, much as Jersild did with words (single) or numbers (double). What the authors found was that the possibility of two rules applying to a single stimulus was particularly difficult for children, less difficult for adolescents, and much less difficult for adults.

The results followed the brain development principles that I discussed in chapter 2. Specifically, not all brain areas showed the same pattern of neural activity across ages. Areas in the parietal cortex, a posterior, more early developing area, showed no real difference in neural activity between children, adolescents, and adults. The supplementary motor area, which is involved in simpler forms of cognitive control, differed in children relative to both adolescents and adults. It was only the prefrontal cortex that revealed a difference between children, adolescents, and adults. Specifically, this area showed a relative reduction across each age group. The results from these studies with task switching confirm earlier neuroanatomical changes that I discussed in chapter 2. Namely, areas of the prefrontal cortex are the last to develop.

As I mentioned in earlier chapters, I studied in Brazil for two years in my early 20s. During the first year and a half, I played tennis two to three times a week. Among the many Brazilians I met, I became very good friends with Plinio Junqueira de Schmidt, a philosophy student who also shared a great interest in playing tennis. One of his favorite lines was "*canso rapido mas acqueco devagar*," which translates roughly to "I tire quickly but I warm up slowly." In other words, he only played well during a short window after a very long warm-up but just before growing fatigued. His humorous line reminds me a lot of the frontal lobes. As we have seen above, the frontal lobes take a long time to develop, not arriving to a fully developed state until a person's 20s. In the next section we will see how this area also starts to show a relatively faster degradation relative to other areas of the brain.

AGING AND THE ERODING PREFRONTAL CORTEX

Aging brings all kinds of changes to the adult body. There is a loss of muscle mass, elasticity of the tendons, and bone strength and mass, and changes in metabolism and in the endurance and recovery of the body. Naturally, the brain shows very similar changes. There is a 5% to 10% loss in brain weight between the ages of 20 and 90. This loss results in a relative flattening of the brain's surface, which can be seen in the wider gyri and sulci, transforming from higher peaks and valleys to smaller hills and canyons. The ventricles that carry cerebrospinal fluid, important for protecting the brain from impact and for lessening the pressure of lying on the skull, begin to grow larger. In short, the brain goes through a series of changes much as the body does.

The changes in brain function across age can be seen in studies that have used MRI (Toga, Thompson, & Sowell, 2006). In an earlier chapter I discussed work done by Arthur Toga and Elizabeth Sowell on brain development. Interestingly, there appear to be clear differences in developmental trajectories depending on which area of the brain we are considering. The temporal lobe tends to show a very flat pattern of change, with this area being relatively stable up until the 60s. The frontal lobe, especially the prefrontal cortex, on the other hand follows the pattern described by my friend Plinio in that it slowly warms up and tires quickly. It tends to show a larger reduction in volume more quickly after age 20 that accelerates even more after age 60. In short, the frontal lobe shows a fast decline.

Along with the neuroanatomical changes in the frontal area that occur with age, there is also considerable evidence of neural activity changes that occur with increased cognitive control demands. Across a series of studies there is evidence that older adults show differences in neural activity relative to younger adults in the frontal and parietal regions that are associated with control. Interestingly, recruitment of these areas has been seen in tasks not normally associated with cognitive control. For example, in tasks that require a person to visually monitor for a stimulus or visual letter detection (Cabeza et al., 2004). A nice example of this is a simple visual detection experiment in which participants are asked to attend to a visual stimulus that flashes once or twice or not at all on a computer screen. The participant is asked to report how many times it blipped. Because eyesight is so acute in younger adults it allows them to do this task in an almost pure sensory mode without much effort. Older adults that have much less reliable sensory modalities are more likely to use control functions to ensure that they have accurately reported the number of blips. The neural activity reported for this task reveals an increase in occipital cortex activity in young adults relative to older adults. Older adults, on the other hand, show an increase in brain activity in the dorsolateral prefrontal cortex. This pattern was even more pronounced when they were asked to remember whether a word presented on a computer screen matched one of four words presented earlier. In short, older adults use cognitive control to regulate the processing of common visual stimuli. This effect is greater in working memory tasks that require a person to keep things in mind for a short period of time.

Another pattern that is often observed in these studies that require cognitive control is the need to resort to compensatory processing. The brain is much like the body. For example, a person who has problems with respiration might show a decrease in blood oxygenation. The body has a good system for detecting changes and will adapt to it. To compensate for this the heart will pump faster in order to get more blood circulating and hence make sure that the blood's oxygen level does not drop too much. In a similar fashion, older adults show compensatory processing in that sensory perception becomes more subject to control relative to younger adults. Another pattern that has been observed is a recruitment of opposite hemisphere homologues relative to younger adults. Hence, the brain activity of older adults will tend to involve both hemispheres to a greater extent relative to young adults.

The fact that older adults show both increased activity of the frontoparietal network and compensatory activity in both the right and left hemisphere also fits in well with results from the task-switching literature. In one study, Jimura and Braver presented a group of young and older adult monolinguals with a task cue followed by a single word on a computer screen (Jimura & Braver, 2011). Participants were instructed to make two different types of decisions. When shown the cue "MAN-NAT," participants had to decide if the word was manmade or natural. When shown the cue "LRG-SML," participants had to decide if the word was larger or smaller than a computer monitor. In one condition, participants were asked to make only one type of decision during a block of trials. In another condition, participants were asked to make different types of decisions for different trials. For example, they might be given the cue "MAN-NAT" for three separately presented words. But then the cue could switch to "LRG-SML" for the next two words. In this mixed condition, participants had to be particularly vigilant as the cue could change at any particular time.

The results suggest that task switching may be more reactive for older adults compared to young adults. Specifically, the cue that signaled a task led to increased activity in both frontal and parietal areas in young adults. This signals a proactive state in which the young adults are anticipating what they will have to do. In older adults, however, only the parietal lobe was responsive to the cue. This suggests that older adults were behaving in a more reactive way. Specifically, they were attaching the cue to a specific response. Young adults were predicting what would need to be done and then allocating resources accordingly. The most interesting part is that both groups showed similar performance, although the older adults were slower overall in raw reaction time. This suggests that older adults may be adapting the way in which their cognitive control systems deal with tasks in order to maintain their behavioral performance.

A subsequent study sought to look at how strongly linked the network is in both younger and older adults. To do this David Madden and Roberto Cabeza, two researchers interested in the brain changes that occur in aging, asked a group of younger and older adults to perform a task similar to the one I mentioned above (Madden et al., 2010). They had to decide if a word was manmade/natural or large/small. The results were compatible to those found in the previous study. The results found differences in the strength of the networks of each group. Specifically, the frontal and parietal areas were strongly linked when processing the cue (MAN-NAT) for young adults. In older adults, this network showed a stronger link when processing the actual word. In short, younger adults were able to predict ahead of time what they would be doing. Older adults seemed to be reacting to the target on each task.

The fact that older adults and young adults show differences in executive function is interesting because it shows that frontal lobe function that changes across age affects tasks that involve executive function. As I noted earlier, frontal lobe development follows a "slow to warm up fast to tire" trajectory. Up until now we have considered the nature of cognitive control as existing mostly outside the realm of language. In the next section we will consider how control relates to language proficiency.

CONTROL AND LANGUAGE

In earlier sections, we considered Piagetian theory and how it relates to cognitive control. Specifically, the ability to represent things in the mind was invariably related to the ability to ignore what was in the environment. The use of symbols is an extension of this ability in that one thing comes to stand for another. With this ability comes language, which calls up all kinds of external and internal events and states without having to act them out.

If, in fact, the neo-Piagetian view that cognitive control is related to language holds, then we should expect that measures of cognitive control would be tightly correlated with measures of language during child development. The relationship between cognitive control and language development has held up in the literature. For example, work with young children has found evidence that the neural signatures associated with integration of information across sensory modalities at a young age differed depending on linguistic performance (Benasich, Gou, Choudhury, & Harris, 2008). Children ages 16, 24, and 36 months were administered a standard battery of language tests that involved having to produce and understand language. The children were then placed in an EEG machine to look at the background physiological activity of the brain. Interestingly, even though the children were not performing any task, the EEG still showed differences in the coherence of brain networks involved in integrating information across senses. The difference in coherence was related to improved language performance. Finally, a subsequent study found that neural differences observed at these early ages also correlated with their ability to produce language at age four (Gou, Choudhury, & Benasich, 2010). Taken together these results suggest that neural activity in frontal brain areas involved in integrating information across senses is related to language development.

The relationship between cognitive control and language processing can also be found in older children. In one study conducted by Elizabeth O'Hare along with Susan Bookheimer and Elizabeth Sowell at UCLA, a group of children, adolescents, and adults were given a verbal task that required them to keep a set of letters in mind for a short period of time (O'Hare, Lu, Houston, Bookheimer, & Sowell, 2008). For example, participants were presented with the letter string LPTCWK for 1.5 seconds, then it was taken away for 3 seconds. Afterward, they were shown a single letter (i.e., *t*) and asked if that letter was present or not in the original string they saw. In addition, the experimenters also varied how many letters were shown in the string in order to create a medium load condition (i.e., L#TC#K) and a low load condition (i.e., ##T###). The results from the memory task revealed no differences across age groups in terms of accuracy. In terms of reaction time, children were slower than adults in general. However, the children did not show a larger slowdown across conditions. Studies generally find that children show slower reaction times relative to young adults. Hence, it is not so surprising that children would be slower overall.

The surprising part of the study was found in the results of the neural activity. Adults and adolescents showed an increase in the neural activity in the entire frontoparietal circuit as the cognitive load increased. However, children showed increased activity, but only in smaller portion of the lateral prefrontal cortex. This is an interesting result for a few reasons. First, it suggests that across age groups similar patterns of performance are maintained even though the neural firing patterns differ considerably. Secondly, it shows a different pattern from that observed in older adults. Hence, children are slower just as older adults are slower than young adults. However, this slowing is not necessarily associated with the same pattern of neural activity.

The fact that children show different patterns of neural activity relative to adults is interesting because it suggests that early in life information is processed differently than it is later in life. In chapters 2 and 3 I discussed a number of studies that show how children are more tied to sensorimotor processing relative to adults. In those two chapters we focused for the most part on the sensorimotor aspect of learning. The results from the studies of cognitive control kind of flip the question on its head. Specifically, if children are using more sensorimotor processing then perhaps they are less tied up in trying to control the process. In short, they may be, like older adults, more reactive than proactive.

Most studies of development focus on the early years of child development. But development doesn't stop at age 12. Adolescence represents a period of intense change. These changes can also be seen in typical cognitive control tasks. For example, Amy Finn and Mark D'Esposito at UC Berkeley asked a group of adolescents to perform a cognitive load task around the age of 14 and then again around the age of 18 (Finn, Sheridan, Kam, Hinshaw, & D'Esposito, 2010). The task asked participants to look at a letter string and then decide if a single letter matched that string or not. Cognitive load was manipulated by presenting strings of 5 (high) or 2 (low) letters. The results revealed faster reaction times for the older adolescents across both conditions of load. There was also an increase in reaction times for the high load conditions across both ages. However, there was no change in the size of the load condition across ages. That is, young adolescents did not show a disproportionately larger slowdown relative to older adolescents in the high load condition. Most importantly, there was no difference in the accuracy across load or age.

Despite there being no difference in the performance on each task, there were still important differences in the brain activity patterns between young and older adolescents. Young adolescents showed stronger brain activity in the lateral prefrontal cortex and the hippocampus in the high load condition relative to the low load condition. However, older adolescents showed increased activity but only in the lateral prefrontal cortex under high load conditions. Hence, young adults show presence of additional brain areas that are not seen in older adolescents.

In previous chapters, I discussed the role that the hippocampus plays in memory. H.M., whose hippocampus was removed, was unable to learn new forms of declarative memory, the kind you can talk about. For him, life was living in the moment. Once something had passed he no longer had conscious memory of it. The fact that young adolescents have increased neural activity in the hippocampus suggests that they were actually forming new memories of the task. Rather than applying existing strategies they were likely consciously thinking through what the strategies might be. Finally, it is possible that the prefrontal activity may also have been related to this different strategy. That is, young adolescents may have been focusing on remembering what the items were rather than simply performing the task. This might imply that young adolescents are still more tied to sensorimotor processing than older adolescents. Hence, even though prefrontal cortex plays a role across all of development it may be playing a different role in adults than in adolescents and children.

ON CONTROL AND DEVELOPMENT

In this chapter, I discussed the nature of cognitive control across the life span. Work in the development literature suggests that the ability to ignore certain types of information in the environment is fundamental for development. I also discussed work in the task switching literature that has evolved somewhat independently of this literature. The work in this literature is best summarized by an expert in the field:

> Every response which an individual makes is in a sense an act of adjustment. Every stimulus that comes to a responding organism is in a sense a test of the adaptability, the power of adjustment, possessed by that organism. The simple reaction time stimulus calls for a simple adjustment; the complex mental task calls for a more complex adjustment. Between the two there is a continuous gradation. When we pass from the isolated simple response to responses of a continuous serial character, we enter a field where the influence of a mental set begins to take form. The mental set is itself an adjustment. It is a response pattern of a given character, called into being by previous stimuli, deriving its strength from previous practice. It initiates preparatory reactions prejudicial to a given type of response. The efficiency of a mental set is expressed in part by the degree to which it wards off irrelevant reactions, in part by the facility with which it effects the adjustment required for each successive response.
>
> Just as a mental set is an adjustment in terms of higher units, so shift is an adjustment called into play by a superimposed mental set or by a forced change brought about by a change in stimuli. The two are inseparable. The uniformity of a mental set, or the high efficiency with which it may operate, does not bespeak the absence of shift. The necessity for shift continues to be operative in the form of constant readjustments even though these may be hidden by the influence of practise or other factors.

This quote is interesting for a few reasons. For one, it places mental set shifting within a context. Rather than being the study of switching between tasks, it is really the study of adjusting to different circumstances. The quote could be referring to the act of driving or to the act of cooking. But there is another aspect of this quote that rings true to me. It seems to also capture some of the aspects of development that I have discussed in this chapter. Like task shifting or switching, development can easily be conceptualized as involving constant adjustment as well. Even between early and late adolescence there are profound changes in the type of information that is processed by the brain. It is also a fundamental difference between our brains and that of every other primate species on the planet. As humans we can adapt to many different circumstances so quickly that we actually ended up changing our habitats. All of this is due to our ability to constantly adjust. By now you may be wondering who wrote these original words. It was none other than Arthur T. Jersild from his original study in 1927 on task-set shifting. As you can see, these words ring true even more than 80 years later, as I am writing this book.

NOTES

1. This fact has been found to be incorrect upon further inspection. It turns out that infants have pretty amazing capabilities at birth. For example, infants have the ability to track faces at birth. But even though Piaget might have gotten the timeline wrong, there are a great number of researchers who still admire the spirit of his view even if the details were not entirely accurate.

7

CONTROL, FLEXIBILITY, AND THE TWO LANGUAGE DILEMMA

IN THE LAST chapter, we considered the notion of cognitive control and its relationship to adjustments. As Jersild noted over 80 years ago, adjustments are critical for human cognition. Interestingly, this notion also characterizes bilingualism very well. Think about the examples of bilingualism we have covered so far. In chapter 3 I discussed work by Nuria Sebastian-Galles and Laura Bosch on bilingual infants that can learn to distinguish two different vowel sets in Catalan and Spanish even though both languages have similar rhythms. In chapter 5 I talked about cases where adoptees seemingly lose the ability to speak their first language even when they are immersed in their second language as late as age eight. I discussed these two examples in light of different theoretical constructs, age of acquisition (AoA) and language proficiency. However, if we step back and think about how these two examples are similar, I think you will quickly realize that both of these examples are a testament to bilinguals' ability to adjust. Even though Jersild was not thinking about bilingualism, the concepts that emerge from the set shifting literature are very characteristic of bilingualism.

I have personally experienced this type of set shifting across my entire life. As an infant, I was exposed to both Spanish and English from birth. I visited Mexico throughout my childhood for long periods of time. At age 20, I studied in Brazil for two years. Finally, at age 34 I embarked on learning German. In each of these instances, my mind and brain had to adapt to change. Many people might fret at having to stuff that many languages in one head. And trust me, I have fretted about it so much that I actually made it into a career. But the biggest insight for all of this occurred at age 20, during a period of initial immersion in Brazilian Portuguese.

In chapter 1, I told you about first few months in Brazil during which I had a period of apparent language loss. During the first three months Portuguese took center stage, and everything else suffered. In some ways, this experience has a thread both in AoA, because I sung in Portuguese as a child and eventually lost my accent, and in proficiency, because it shows how immersion and use led to rapid improvement in my spoken language ability. Portuguese lay dormant in me waiting until the moment of immersion at which point it took on a life of its own. My experience in Brazil also has a third component that relates to the topic in our current chapter: cognitive control. I know now that I had not really lost my first two languages. Eventually, I regained both English and Spanish skills over time. Whatever I had gone through must have involved a temporary change, it was the product of an abrupt shift in my environment that required profound adjustments.

Work in the bilingual literature has identified the cognitive mechanisms that are at the heart of these immersion effects. Jared Linck, Karen Sunderman, and Judith Kroll at Penn State University wanted to get to the bottom of this effect (Linck, Kroll, & Sunderman, 2009). To do this they compared two separate groups of adult second language learners. One group had learned in a classroom setting, and the second group were abroad at the time of testing. Both groups were matched in their second language knowledge and the amount of time they had spent studying the second language. Hence, there was no difference in the quantity of language exposure. The differences were in the quality of this exposure. Participants were given a set of words that were related to each other or not. In addition, they were also given a category cue (e.g., fruit) and asked to produce as many items from that category. The results revealed key differences between those who were immersed and those who were not. Participants immersed in Spanish produced fewer items in English and more items in Spanish than classroom learners. They also showed less sensitivity to translation and category relationships between the two languages.

In chapter 5, we discussed Kroll's model, which suggests links between a bilingual's two word stores and a single store for concepts. What immersion appears to have done was to strengthen those connections between the second language words and their corresponding concepts. At the same time the fact that second language learners were in Spain seems to have dampened the direct connections between the two languages. Kroll and colleagues suggest that this dampening has to do with inhibitory processing. These results also resonate well with theories of retrieval failure such as those proposed by Robert Bjork and colleagues (Bjork & Benjamin, 2011). In this view, memory must involve activation of what is most relevant to a person at a particular moment in time. Remembering everything all the time would be catastrophic. Along the same lines, immersion in adult second language learners appears to result in a shift toward a second language processing mode. During this time, bilinguals essentially lose some access to their first language. So what I experienced in Brazil was not so strange after all. Except in my case I lost access to both of my native languages, with Spanish taking the biggest hit.

The sudden loss in the ability to process a known language suggests a fluidity in the use of languages that could be seen as having to do more directly with language proficiency. I have no doubt that continued use of a language would lead to eventual changes in proficiency. However, this proficiency is also accompanied by an increased need to control that language. This notion of shifting is very similar to the one that Pötzl used many years ago when considering patients who appeared to be in stuck in a language because of brain damage. In the next section, I trace out the natural extension of these concepts. I will show how cognitive control plays a role in language proficiency. I will also review newer literature that suggests that the control needed by bilinguals to manage two languages bleeds out to other types of nonlinguistic control tasks. But first let's go back to the idea of shifting. You see it was in task switching using bilinguals' two languages that the importance of control was first established.

THE LANGUAGE SWITCH

The need to control which language is active has been the subject of considerable attention in the psychological literature. Researchers wondered how bilinguals could keep information in one language from constantly interfering with processing of the other language. The potential for interference is (at least in theory) massive, particularly in view of the overlap in neural tissue. Interestingly, the notion of a switchboard was originally proposed by Wilder Penfield, who felt that there must be some interface between motor and sensory systems in the brain (Penfield, 1950). Penfield then extended this idea by proposing a language switch in order to account for the lack of interference in bilinguals (Penfield & Roberts, 1959).

As the 1960s approached, more and more researchers sought to identify language switch costs. However, results from a series of studies found mixed evidence for effects of language switching on behavioral performance in healthy young adult bilinguals. Some found no evidence of language switching effects (Dalrymple-Alford & Aamiry, 1969; Kolers, 1966), while others found significant effects of language switching particularly in terms of speed of processing (MacNamara, Krauthammer, & Bolgar, 1968; MacNamara & Kushnir, 1971; Soares & Grosjean, 1984).

A second wave of interest in language switching came during the late 1980s through the early 2000s. With the advent of computer controlled experiments, researchers could now ask participants to look at the responses that participants made to one item at a time. In a classic study, Grainger and Beauvillain asked participants to make word decisions to a set of French and English words. The catch was that the words could be all in French, all in English, or they could be in either language (Grainger & Beauvillain, 1988). Bilinguals were slower to make word decisions when the language was mixed relative to the single language conditions. In other words, having to read words in both languages made bilinguals slow down.

Von Studnitz and Green expanded on these results by asking participants to make two different types of word decisions for stimuli in paired alternation between German and English (von Studnitz & Green, 1997). This paired alternation (GGEEGGEEGGEE) allowed the investigators to look at the difference between a switch trial (GG**E**E) and a no switch trial (GGE**E**). In Experiment 1, participants saw words in either a green or red color font and were asked to decide if the word was a real word or not. The color indicated the language of their word decisions. If the word was in red, they would decide if it was an English word or not. If it was in green, they would decide if it was a German word or not. Half of the subjects paired English with red and half of them paired German with red. Experiment 2 allowed participants to make word decisions without having to attend to any particular language.

The results revealed slower responses when participants had to make a word decision in only one language. This is not surprising, since this required bilinguals to focus on one language while ignoring the other. The additional control needed to focus on only one language for each word would naturally lead to slower responses. The interesting result was that both experiments showed slower responses for the switch condition (GG**E**E) relative to the no switch condition (GGE**E**). Hence, it appears that having to go from one language to the other slows people down even if it has no role in the response that they are making.

My incursion into this literature came from a slightly different direction (A. E. Hernandez et al., 1996). In one of my first experiments in graduate school, I asked bilingual participants to listen to a spoken paragraph that had gaps in it. In these gaps, I would present written words on a computer screen either in the same language or in another language. It was an experimental twist on a task made famous by David Swinney and Donald Foss, the former a member of my doctoral committee and the latter a future colleague (Foss & Speer, 1983; Swinney, 1979). My idea was to manipulate the predictability of the language of the word in that particular paragraph. In the mixed language condition, the visually presented words that filled gaps in a spoken passage could appear in either language. In the blocked condition, the word would appear either in the same language or in the other language during a particular passage. Hence, subjects in the mixed condition were not sure of the language in which the visual word would appear. When the language of the auditory paragraph and word matched, subjects processed the meaning of the word in both the blocked and mixed condition. When the language was completely predictable the meaning affected reading speed. Interestingly, studies with EEG suggest that a sentence that ends with a word in the other language leads to an elongated N400, the negative going wave that is sensitive to how easily a word can be read (Moreno, Federmeier, & Kutas, 2002). Hence, it appears that changing languages involves having to read a word in a more effortful manner.

The most interesting effect came when participants were not sure of the language in which the visual words would appear. In this unpredictable or mixed language condition, participants were able to read the words as fast as they could in the predictable condition. They were also just as accurate. In terms of meaning, there was an effect when the language of the sentence matched the language of the word. However, when the languages did not match, there was no effect of meaning. So bilinguals seemed to be trading off the processing of meaning in order to complete the task successfully and read the words on the computer screen.

The nature of switching has also been explored in production tasks. In a classic study, Meuter and Allport asked participants to say the names of single digits (1–9) out loud (Meuter & Allport, 1999). Participants were presented with single numbers in either yellow or blue. Each person was trained to name the single digit in one language or the other depending on the color of the font. The results revealed that participants were faster for switch (GG**E**E) conditions relative to no switch conditions (GGE**E**). Furthermore, there was an effect of mixing such that naming digits in a single language was faster than having to switch between languages. These results confirmed that switching slows people down even when saying digits out loud. Since digits are easily named in both languages, the results suggest that bilinguals are slowed by the act of switching and not by a lack of knowledge in either language. Finally, both Meuter and Allport (1999) and Jackson et al. (2001) have found that switching is asymmetric such that it is harder to suppress the more proficient language. This suggests that added control is present in switching and particularly needed when having to turn off the more proficient language and turn on the least proficient one.

The results we have discussed so far fit in with the view that bilinguals slow their responses during switching due to the need to control which language will be accessed. As I noted last chapter, switching tasks are known to be reliant on frontal lobe function. This carries two implications. First, there should be brain activity in areas of the frontal lobe when looking at conditions of switching relative to conditions with only one language. Second, there should be changes in the size of this switching effect depending on the age of the subjects tested. Specifically, children and older adults should show larger switching effects relative to young adults due to differences in frontal lobe function due to age.

Looking at the changes across age in switching for a bilingual language task was a topic of great interest for us. Liz Bates had become interested in picture naming and was intrigued by the possibility of looking at how each language interfered with the other. Along with Kathryn Kohnert, a grad student working with Liz Bates at the time, we set out to design a language-switching task with pictures. In our task, we would ask bilinguals to name a series of pictures using only one language and compare that to a condition where they had to name some pictures in Spanish and some in English. The key was to not let people know which language they had to use until the moment they saw the picture. When I mentioned this idea to Liz Bates she suggested that we playback over a set of speakers the word *say* or *diga* before each picture so that a subject would know what language to use when saying the name of that picture.

In 1996 we began to gather data with this task. Over the next two years, Kathryn Kohnert and I collected data from over 30 older adult and 100 child bilinguals. The results showed a clear developmental pattern (A. E. Hernandez & Kohnert, 1999; Kohnert et al., 1999). Children were much slower when switching between languages compared to naming pictures in one language. The size of this switching effect got smaller in adulthood and stood stable through bilinguals' 30s and 40s. However, for older adults who were 60 and above the size of the switching effect got considerably larger, reaching half a second relative to the 1/10 of a second in college-age adults.

Our results were eerily similar to those that I reported in the previous chapter involving task switching across different age groups. In studies of task switching, the general finding is that children and older adults show larger switching effects relative to young adults (Crone, Bunge, van der Molen, & Ridderinkhof, 2006; Verhaeghen et al., 2005). Along similar lines, work in my laboratory has found that bilingual children and older adults show larger language switching effects. These effects seemingly track the wax and wane of the prefrontal cortex. The question is whether language switching would also be revealed in the amount of brain activity seen in the prefrontal areas of the brain.

IN SEARCH OF THE LANGUAGE SWITCH

The interest in language switching had also extended to the possible neural locus of this effect. At the neuroanatomical level, researchers had found mixed evidence for one area that was exclusively involved in language switching. Classic cases in aphasia presented by Pötzl and Kauders suggested that the language switch was localized in the supramarginal gyrus, which is in the parietal lobe (Herschmann & Pötzl, 1983; Kauders, 1983; Pötzl, 1983). However, others had found evidence of patients with lesions in the supramarginal gyrus area, the posterior part of the sylvian fissure, and adjoining areas of the parietal lobe in whom there was no switching difficulty (Gloning & Gloning, 1983; Minkowski, 1983; Stengel & Zelmanowicz, 1933). One of the cases presented by Stengel and Zelmanowicz is of particular interest, given that it was one of the few cases in which language mixing appears both in spontaneous conversation and in picture naming for a person with a motor aphasia that was thought to be due to a frontal lesion. The most important point is that there was very little consistency in the areas that when damaged were leading to difficulties with language switching. In short, work in the neuropsychology literature up until that point had not found any evidence of a clear neural center involved in language switching. Rather there was evidence of lesions to frontal, temporal, and parietal areas that led to problems with language mixing. Given these findings there appeared to be no specific area of the brain that was dedicated exclusively to language switching.

The brain areas involved in language switching have also been explored in a series of studies using modern neuroimaging techniques. The first of these studies were published almost simultaneously by our group and by Cathy Price, David Green, and Roswitha von Studnitz, (Price, Green, & Von Studnitz, 1999). Price, Green, and Von Studnitz looked at the nature of translation and language switching in a group of German-English bilinguals. Participants were asked to read a word on a computer screen and mouth the word or its translation while being scanned with positron emission tomography, a measure of blood flow in the brain. Translation relative to reading led to increases in subcortical areas involved in motor control such as the putamen and the caudate nucleus as well as other areas involved in control such as the anterior insula, cerebellum, and the anterior cingulate gyrus. In addition, Price and colleagues asked participants either to respond in one language or to alternate between languages when responding in a particular block. When comparing switching to a single language, there was increased activity in the supramarginal gyrus and inferior frontal gyrus for switching relative to the single language condition. Interestingly, there was no increased activity in the dorsolateral prefrontal cortex in either the translation or the switching condition. Hence, in that study there appeared to be little need for the type of higher order control that I discussed last chapter when I reviewed the monolingual task switching literature.

The nature of language switching that we had explored both with older adults and children was also the topic of the first two neuroimaging studies conducted in my laboratory (A. E. Hernandez, Dapretto, Mazziotta, & Bookheimer, 2001; A. E. Hernandez, Martinez, & Kohnert, 2000). In these studies, we asked a group of Spanish-English bilinguals to imagine saying the names of a set of pictures "in their heads." This covert naming task was used extensively at the time to avoid the head motion that accompanies talking thereby making our images more noisy. The results revealed more activity in the dorsolateral prefrontal cortex when bilinguals had to switch between languages compared to when they had to simply say the names to themselves in only one language. A second study conducted with Susan Bookheimer, Mirella Dapretto, and John Mazziotta at UCLA replicated the switching effect that I had found in my original experiment. Specifically, there was more neural activity in the right dorsolateral prefrontal cortex in the mixed language condition relative to the single language condition. Interestingly, the reverse comparison revealed increased activity in Broca's area (which is part of the ventrolateral prefrontal cortex) on the left. Taken together these results suggest that having to switch between languages involves activity in brain areas involved in cognitive control.

The importance of the dorsolateral prefrontal cortex (DLPFC) in language switching has also been observed in a series of studies that have investigated the effects of transcranial magnetic stimulation, or TMS. TMS uses a small magnetic pulse via a coil applied to the surface of the scalp. This pulse penetrates the surface of the cortex and can temporarily alter firing of neurons. In one study, Holtzheimer and colleagues (Holtzheimer, Fawaz, Wilson, & Avery, 2005), applied a TMS pulse over the left DLPFC to an English-German and an English-Spanish bilingual who were being treated for major depressive disorders. Both patients reported thinking in their second language to a greater extent after the treatment. The fact that stimulation of the prefrontal cortex leads to a language invading a patient's thought to a greater extent provides an indirect link between the dorsolateral prefrontal cortex and control of a language.

Since the original studies, a number of additional studies have looked at the nature of language switching. These studies have identified a much larger set of structures that are involved in the control of language. For example, Crinion and colleagues asked participants to look at pairs of words that were related or not and that matched across languages or did not (Crinion et al., 2006). Results showed a decrease in neural activity in the anterior portions of the temporal lobe when participants read words that were related to each other. In addition, activity in the caudate nucleus differed depending on whether the language matched or did not. As we noted earlier, the caudate nucleus is known to play a role in the learning of new tasks with rules that can be stated verbally and is highly linked with activity in language areas of the temporal and parietal cortices along with the dorsolateral portion of the prefrontal cortex (Ashby & Crossley, 2012). Angela Friederici, one of the directors of the Max Planck Institute for Mind and Brain in Leipzig, Germany, has proposed that the caudate nucleus plays a role in language control generally speaking and may play a particularly crucial role for bilinguals, who must maintain control of each language (Friederici, 2006). Finally, the importance of the caudate nucleus in control has also been documented in the neuropsychological literature. In a case study,

Abutalebi and colleagues tested a trilingual patient with damage to the caudate nucleus (Vitali et al., 2010). The patient had no difficulty in naming pictures but showed involuntary switching during language production during regular speech. Thus the caudate nucleus seems to play a role in the control of language as part of a frontal-subcortical loop that is dedicated to the use of declarative memory and cognitive control. Finally, Gabrielle Garbin, Albert Costa, and colleagues have found that the caudate nucleus plays a role in switching to the dominant language in highly proficient bilinguals (Garbin et al., 2011). Thus the caudate nucleus is important in overcoming interference for bilinguals.

The nature of language switching has received considerable attention in the literature (Hervais-Adelman, Moser-Mercer, & Golestani, 2011). These studies have begun to converge on a network of brain areas; the dorsolateral prefrontal cortex, the caudate nucleus, the anterior cingulate gyrus, and the supramarginal gyrus in the parietal lobe play important roles in control tasks for monolinguals. Many of these areas may be recognizable to you from the previous chapter on control in bilinguals.

To make sense of these data, Jubin Abutalebi and David Green have looked at how each component of the network plays a role in language switching and more generally in control (Abutalebi, Della Rosa, Tettamanti, Green, & Cappa, 2009; Abutalebi & Green, 2007; Abutalebi & Green, 2008; Green & Abutalebi, 2008). (See figure 7.1.) They postulate that areas that are more anterior, such as the dorsolateral prefrontal cortex and the anterior cingulate gyrus, are involved in selecting

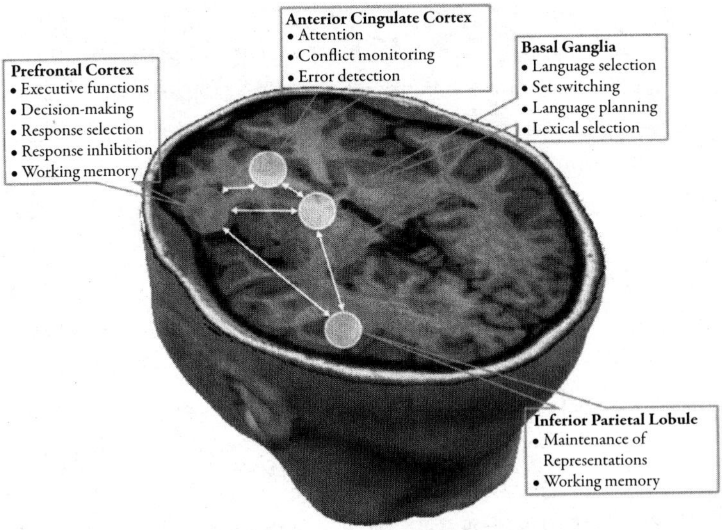

FIGURE 7.1 Areas of the brain involved in language control. (Reprinted from Abutalebi, J., & Green, D. Bilingual language production: The neurocognition of language representation and control. *Journal of Neurolinguistics, 20,* 242–275. Copyright [2007], with permission from Elsevier).

the appropriate language for response and correcting for any possible errors in this process. The caudate nucleus, which is part of the basal ganglia, plays a role in control and is thought to be involved in supervising the correct selection. Finally, the supramarginal gyrus, in the parietal lobe, is involved in providing a bias toward the correct language and away from the incorrect language.

One way to envision this is as a wave of brain activity across these brain areas that leads to the correct selection of a language. The supramarginal gyrus, in the parietal lobe, would be initially involved in providing a bias toward one of the languages. Slowly activity in the anterior cingulate gyrus and the caudate nucleus would begin to refine these biases to a greater extent. In the final step, the prefrontal cortex would become more active as the correct language is selected.

However, this control network is not important merely for bilingual language switching. Abutalebi and Green also argue that this network is used when a second language is spoken with lower proficiency. In chapters 4 and 5, I considered that proficiency could be conceptualized as having to do with efficiency. The idea is that expertise in a language involves having to do things well. When we are less proficient, it requires more effort to obtain the same result. Naturally, control areas should become involved in any new activity we take on. I also discussed research from my colleague Sian Beilock, who has found that when consciously thinking about their swings novice golfers actually do better relative to expert golfers. The same principle should apply when comparing a less proficient to a more proficient language within a bilingual. That is, there should be a greater need for control mechanisms when a second less proficient language is produced.

Work in my laboratory has confirmed this in a cued picture naming experiment on a group of late Spanish-English bilinguals. The results from this study using functional magnetic resonance imaging (fMRI) revealed increased activity in four separate areas: the dorsolateral prefrontal cortex, the anterior cingulate gyrus, the anterior insula, and the fusiform gyrus. The dorsolateral prefrontal cortex and anterior cingulate gyrus, as I noted earlier, are involved in selecting the correct language. The anterior insula is known to play a role in the articulation or physical movements necessary for producing words. Finally, the fusiform gyrus, as I noted in chapter 4 when talking about the perceptual expertise needed to recognize objects, is present since the task involves having to identify the labels associated with an object.

The results from the fMRI study conducted in my lab provide an interesting extension of Abutalebi and Green's model. They suggest that control areas do not operate in a vacuum. At least in our study, we found that frontal control areas were used to activate both the object representations and the motor programs associated with those objects. In short, speaking in a late learned second language requires bilinguals to cycle between what an object is and its corresponding label. This result is consonant with Judith Kroll's idea of inhibition that we discussed earlier. However, in our case the subjects were immersed in their second language but had attained a higher degree of proficiency. This suggests that once the initial adaptation to a second language occurs, late bilinguals will need inhibition in order to overcome interference from their more proficient first language.

Based on the data I have discussed we should be able to observe similar control networks in a group of bilinguals performing a nonverbal switch task. Surprisingly, we did not find such an effect in 2001 (A. E. Hernandez et al., 2001). At the time I was a bit disappointed with these findings. I was really hoping that we could find some parallels or differences between language switching in a bilingual and monolingual design. However, in retrospect the lack of a within-language switching effect had two possible causes. One possibility is that we had very little sensitivity to detect this difference. That is, there was an effect but we missed it. Imagine that I turned on the light in your house on a very sunny day in the middle of July. You might not even notice it was on. However, late at night you would be able to see the light very easily. Just like seeing a light, the fact that nothing shows up when we compare two conditions in an experiment, does not mean that there is no difference. It just might be that our measure is not sensitive enough to detect these differences. The second more interesting possibility is that bilinguals are not as sensitive to these task manipulations as monolinguals are. That is, I observed no within-language switching effect because bilinguals are very good switchers. At the time I had read the work of Ellen Bialystok with attentional flexibility in bilingual children (Bialystok, 1988; Bialystok & Harris, 1992). What I did not know was how this profound idea would influence the field in the years to come.

THE BILINGUAL ADVANTAGE

Ellen Bialystok began to suggest in the late 1990s that bilingual flexibility might actually enhance thinking in nonlanguage functions (Bialystok, 1999). In two separate studies, Bialystok asked children to do a dimensional card sort task in which children are given a set of cards with colored shapes ($\square_{\text{red}}$ $\bullet_{\text{green}}$ $\bigcirc_{\text{red}}$ $\blacksquare_{\text{green}}$). At first, children are asked to sort the cards by color ($\square_{\text{red}}$ $\bigcirc_{\text{red}}$ / $\blacksquare_{\text{green}}$ $\bullet_{\text{green}}$). Once they have finished creating different piles for each color, they are asked to sort these again by shape ($\square_{\text{red}}$ $\blacksquare_{\text{green}}$ / $\bigcirc_{\text{red}}$ $\bullet_{\text{green}}$). This task shares some aspects of an executive task. Hence, what is difficult for children is that on the second step they need to ignore the color and attend only to the shape. Bialystok found that bilingual children outperformed monolingual children on this task.

There are a few possible reasons that monolingual or bilingual children might differ on these tasks. Bilinguals may differ from monolinguals in their ability to conceptualize what color and shape are, in their ability to change their motor responses for each task, or in their ability to shift from one rule to another. To test this, Bialystok and Martin (Bialystok & Martin, 2004) asked children to perform a series of tasks that probed the concepts, the motor response, or the act of switching independently. What they found was that bilingual and monolingual children did not differ in their ability to represent each rule. They were also equally good at switching their motor responses. The main difference was in their ability to switch between different concepts. In short, bilingual children were better at ignoring one dimension and attending to a different one.

Follow-up work by Hanako Yoshida and her graduate student Crystal Tran has found not only that these advantages appear early in life but also a possible cause for the advantage in bilinguals. In a first study, Yoshida and Tran asked a group of young children to play the Hungry Fish game, a cognitive control task designed for children (Yoshida, Tran, Benitez, & Kuwabara, 2011). In adult tasks, like the ones I discussed in the last chapter, the usual instructions are to ask the person to indicate some aspect of the stimulus. For example, they might ask which way the middle arrow is facing in the following set of stimuli (→ → → → →). To look at cognitive control, experimenters create a congruent condition (→ → → → → or ← ← ← ← ←) or an incongruent condition (→ → ← → → or ← ← → ← ←). The task is simply to indicate which way the middle arrow is facing. To make finding the middle arrow a bit easier a small asterisk is flashed above the location where the middle arrow will be immediately before presenting it.

Although this task seems a bit complicated, most adults are able to learn to do it quite quickly. However, children find this task too complex. In order to simplify this, Rueda et al. (Rueda et al., 2004) designed the hungry fish game (Figure 7.2). In this task, children are shown a row of fish and asked to feed the middle fish that is hungry by pointing to its mouth. Children are then given feedback with a "woo-hoo!" if they are correct or a "huh?" if they are wrong. The results from this task show that children improve in their ability to attend to the fish and perform the task in both accuracy and speed even after the age of 10. Hence, this is a good task for measuring the development of cognitive control.

Yoshida, Tran, and colleagues were interested in testing even younger children. So they presented a simplified version of the hungry fish task to make it friendly for young children. The results yielded an interesting pattern of results. In the hungry fish game, bilingual children were faster and more accurate than monolinguals. Interestingly, this advantage appeared at age 2.5 years but then became smaller between the ages of 3 and 4. Bilingual children showed a short-lived advantage in their ability to attend to the relevant information (the middle fish) and ignore the irrelevant information (the distracting fish around the hungry fish). This is evidence of a bilingual advantage early in life that fits in nicely with other findings by Bialystok and colleagues.

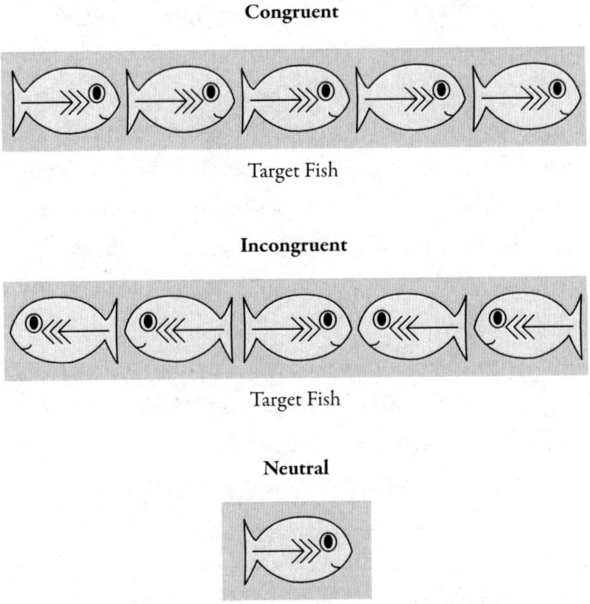

FIGURE 7.2 A set of stimuli used in the Attention Network Test for three types of trials. (Reprinted from Rueda, M. R., Fan, J., McCandliss, B. D., Halparin, J. D., Gruber, D. B., Lercari, L. P., et al. Development of attentional networks in childhood. *Neuropsychologia, 42,* 1029–1040. Copyright [2004], with permission from Elsevier).

THE BILINGUAL DISADVANTAGE

The fact that bilingual children are better at ignoring certain concepts and attending to others is interesting in light of findings in my own laboratory. Work we had done on language switching had found that both children and older adults show larger language switching effects than young adults. In addition, we had traced the brain activity in young adults in this condition to increased activity in the dorsolateral prefrontal cortex. At the time I worried about the implications of my study. After all, someone could argue that bilinguals get confused between their languages and fall apart in older adulthood. Or maybe, they could argue that bilinguals may have to learn twice as much and that they have to work twice as hard.

In fact, work in the bilingual literature has found some evidence of a bilingual disadvantage in the domain of word production. For example, bilinguals are slower to say the names of pictures in either language relative to monolinguals (Ivanova & Costa, 2008). Work by Tamar Gollan, at University of California, San Diego, has confirmed this (Gollan et al., 2008; Gollan et al., 2005). Gollan reasoned that if bilinguals have to deal with twice as many labels then they should show more difficulties with having to come up with the specific word in one of their two languages. And, not surprisingly, they do. Bilinguals show larger tip-of-the-tongue (TOT) states than monolinguals.

TOTs are actually common even in monolinguals. One of the most common types occurs with proper names. When we moved to Texas in 2003, I remember meeting a father of one of my daughter's classmates during a Christmas party in her classroom. We engaged in a bit of small talk for about 10 minutes or so. The part that stuck out at me was how he repeated my name a few times during the conversation. Of course, most of the time I don't worry about the name but usually get caught up in the details of the conversation. At the end, he said goodbye to me using my first name. When I ran into him a few weeks later, he magically remembered my name. And I... Well, I remembered some of the details of our conversation but not his name. When we stop to remember people's names and fail to get it right we are caught in a classic TOT or tip-of-the-tongue state.

Many times these TOT states are not complete failures in figuring out the name of an item. Sometimes people can actually know a lot about the item they are trying to remember. In the summer of 2008, we were driving to Sevilla after having spent eight days in southeastern Portugal. As we drove we heard a song by a band from the1980s. It was a ballad by a group famous for rock songs. I recognized the band and was trying to figure out its name. I told my wife, "They had that song. Love you like a hurricane." After a couple of minutes I said, "They start with an S... Something like the Spiders." A minute later the announcer came on and said the name of the group: The Scorpions. I knew a lot about the word just not the exact label.

Gollan has looked at a similar phenomenon in bilinguals by exploring factors that influence how bilinguals recognize and produce the names of pictures and compare the response times to those of monolinguals. When bilinguals were asked to classify the objects as either human made or natural, bilinguals and monolinguals showed no differences. However, when asked to produce the names of these pictures bilinguals were slower than monolinguals. After naming the pictures, bilinguals were asked which of these words they could translate. Then Gollan checked to see if the words that could be translated were faster or slower than those that were not. Interestingly, it turns out that bilinguals were faster to name a picture in English when they knew the translation of the word in Spanish. This effect did not appear in monolinguals indicating that there was no difference in the pictures themselves.

To account for these effects, Gollan suggests that bilinguals suffer from greater difficulties coming up with a specific label because of the increased number of words across both languages. The presence of all these words leads to weaker links between the words and their concepts. These effects are not specific to bilinguals. In monolinguals, weaker links are observed in words that occur less frequently. For example, a group of monolinguals will be able to say the word "window" faster when looking at a picture of it than they will be able to say the name of a picture of a squirrel. Now imagine a German-English bilingual who has two concepts but now has four words.

"Window" and "*fenster*" have roughly half the strength that they would in a monolingual. Because these words are encountered frequently you might not really notice the effect so much. However, "squirrel" and "*Eichhörnchen*" will take a much bigger hit because of the fact that these links are weak to begin with in a single language. The poor words get even poorer in bilinguals. The effect is particularly striking in a language spoken with lower proficiency. Gollan suggests that a weaker language behaves similarly to weaker words in one language. Hence, the increased difficulty coming up with a specific word has the same underlying principle in bilinguals and monolinguals.

The work by Gollan just added to my initial fears. Now there was evidence that bilinguals got their languages confused even when they were not switching between languages. In 2004 my initial misgivings were allayed by Ellen Bialystok and Fergus Craik (an expert on the cognitive changes associated with aging), who began to explore the possibility of enhanced cognition in a series of studies with older adults. Specifically, they wondered whether older adult bilinguals might have an advantage in tasks requiring controlled attention (Bialystok, Craik, Klein, & Viswanathan, 2004). Could bilingualism actually reduce changes associated with aging?

The results from Bialystok and Craik's study were stunning. Bilinguals were given the Simon task, in which they had to attend to the color of an object and press a button on the right or left. This is a classic control task, since the color is mapped to the side of response. For example, seeing a red diamond on the right (--- ♦) would require a right-button response. However, the same shape on the left (♦ ---) would still require a right-button response. In this condition there is a mismatch between color and side of response, which leads to slower reaction times. This condition (♦ ---) leads to slower responses than when the color and side match (--- ♦). Bilinguals in their 40s were half a second faster than monolinguals when ignoring the side and attending to the shape. By the time bilinguals reach their 60s this effect morphed into a 1½-second advantage over their monolingual peers. In short, a small advantage in middle age results in a very big advantage in the elderly.

As if an advantage on the Simon task in older adult bilinguals was not enough, Bialystok and Craik carried it even further. Together they wondered if bilingualism might protect against the biggest contributor to mental aging, Alzheimer's disease. Autopsies of the normal population finds that the rate of Alzheimer's disease–like pathology without symptoms can be as high as 30 percent (Valenzuela & Sachdev, 2006a, 2006b). Recent work in the Alzheimer's literature has begun to consider the notion of cognitive reserve. That is, people who otherwise would show symptoms of dementia may be relatively unsymptomatic. The natural question is whether the use of two languages might have a similar type of effect. To determine this, the authors looked through 184 case studies. Of those studied, 51 percent were bilingual. In general, they also found that bilinguals showed symptoms four years later than monolinguals.

One question is whether the bilingual advantage was an immigrant advantage. As I'll discuss below there is a growing set of studies that have failed to find a clear cognitive control advantage for young adult bilinguals. One possible factor that has been considered is whether the bilingual advantage may really be an immigrant advantage in disguise. To control for this possible factor, Bialystok and Craik conducted a second study in which they looked at the effects of immigration status on the bilingual advantage. A second group of bilingual older adults diagnosed with probable Alzheimer's disease were tested. In this case, additional language history information was gathered. Results from this study found that bilinguals were diagnosed with Alzheimer's four years later than monolinguals and showed significant cognitive decline five years later. The effect of immigration status did not in anyway alter this effect. In short, the bilingual advantage, at least in older adults, cannot be explained as an immigration effect in disguise.

The results from Bialystok and Craik's study show that my fear was based on the wrong premise. Initial studies conducted in my laboratory involved a comparison between young and older adult bilinguals. The older adult bilinguals show a large language switching effect relative to young bilinguals. The effect is strong enough to see without any kind of sophisticated response apparatus. In fact, I could feel myself slowing down when I tried the task out. So maybe older adults are a half second slower than young bilinguals when switching. But in the end this cost leads to a huge benefit in that bilingual older adults outperform their monolingual peers. In fact, these results suggest that bilingualism may be a very simple form of mental gymnastics. All the work that bilinguals go through on a daily basis adds up over time. By the time they reach their 60s these advantages are evident in a task that has nothing to do with language. Most importantly, thanks to Bialystok and Craik I no longer have to fear I am the one to expose how bad bilingualism is for older adults. Having to endure these costs may in fact help bilinguals in older age.

A COUPLE OF CAVEATS

Despite the widespread interest in the bilingual advantage, there are studies that have begun to qualify the circumstances under which the effect is and is not observed. There are a couple of issues that need to be considered in this respect. One interesting factor is whether the advantage for bilinguals with two spoken languages also appears in those who use a spoken and a signed language. An initial study by Karen Emmorey, Ellen Bialystok, and colleagues (Emmorey, Luk, Pyers, & Bialystok, 2008) compared a group of monolinguals with two groups of bilinguals who either spoke two auditory languages or spoke one auditory language and a sign language.

Participants were asked to perform a flanker task in which the head of an arrow (e.g., < or >) was surrounded by a group of *X*'s on either side of a set of additional arrows, similar to an adult version of the hungry fish task. Participants were asked to indicate where the arrow was pointing when surrounded by *X*'s or to withhold the response when surrounded by other arrows. The results revealed the previously observed advantage for bilinguals who speak two languages. However, there was no advantage for bimodal bilinguals.

The absence of an advantage is not unique to bimodal bilinguals. For example, recent work by Shanna Kousaie, Natalie Phillips, and colleagues has found similar results with adult French-English bilinguals (Kousaie & Phillips, 2012). In this study, a group of bilingual and monolingual subjects from the Montreal area were asked to perform three separate executive function tasks while being monitored with EEG equipment. Results from this study showed no difference between monolinguals and bilinguals in reaction times and percent correct. Results from the EEG experiment found very subtle differences between monolinguals and bilinguals on the three different ERP components: the N200, an early negative going wave that indicates conflict monitoring, the P300, and the error related negativity that appears when monitoring for errors. However, these three EEG wave differences did not appear across all tasks. This suggests subtle differences between the groups but no clear advantage for bilinguals over monolinguals.

So we are left wondering why some studies find a bilingual advantage and others do not. Part of the answer may have to do with the type of bilingual tested. For example, Emmorey, Bialystok, and colleagues claim that bimodal bilinguals, who are bilingual in a spoken and signed language, do not show the same advantage on tasks of cognitive control due to the way in which both languages are used. Bilinguals in two spoken languages in general tend to have a conflict with regard to which language comes in and out. Because both languages share the same input and output motor and sensory systems there is a conflict between them. Hence, speaking two auditory languages requires control in order to avoid interference between them.

The conflict between auditory and motor programs reminds me of a story that a female colleague of mine (whose name I'll protect out of courtesy) used to tell me. As a teenager my colleague and a boyfriend would often ride in the back seat. When this happened her mom would ask them to clap and whistle at the same time. The idea is that by keeping their lips and hands occupied she would avoid any other possible uses of these two body parts. In a similar vein, unimodal bilinguals who speak two spoken languages would not be able to speak or comprehend both languages at the same time. Lips (and hands!) can only move in one direction at any given moment.

The reason that bimodal bilinguals do not show the same type of interference as that seen in unimodal bilinguals is because of the way in which both languages are used. Because one language uses hands and the other uses a speaker's mouth, bimodal bilinguals end up mouthing the words as they sign. This code-blending is much less conflictive than the type of conflict seen in unimodal bilingualism. Imagine that my colleague's mother had asked her to whistle or clap. Then she might have been able to trick her mother.

So what about the bilinguals in Kousaie's study? Interestingly, these participants are from Montreal, a place where there is frequent use of both languages in daily life. It is possible that this particular context yields less conflict than that seen in an immigrant context, where each language is associated with very specific home and environmental situations. As I noted above, studies by Bialystok and Craik suggest that the bilingual advantage in older adults is not just applicable in immigrant populations. Future work is needed to investigate this further.

Finally, I want to end by considering a couple of studies that have questioned the nature of the bilingual effect. In a meta-analysis that pooled results across many studies, Hilchey and Klein (2011) found considerably more studies that failed to show a bilingual advantage than studies that did find an advantage. Furthermore, they argued that the advantage was present in an overall improvement in reaction time rather than a tendency toward a smaller interference effect. The lack of replication across studies brings up the question of whether there might be underlying differences that are driving this effect. One such difference is socioeconomic status, as studies that control for this factor find that the bilingual advantage disappears (J. B. Morton & Harper, 2007). Finally, studies by Julia Festman, Antoni Rodriguez-Fornells, and Tomas Munte (Festman & Munte, 2012; Festman, Rodriguez-Fornells, & Munte, 2010; Rodriguez-Fornells, Kramer, Lorenzo-Seva, Festman, & Munte, 2011) have found that bilinguals vary in the amount of control that they can use. This research team has documented in a series of papers that bilinguals vary in their ability to produce the name of a picture out loud in the correct language when switching between languages. This difference is related to the ability to perform nonverbal control tasks such as the Simon task I described earlier, in which participants had to press the right button if the item was red. This task creates greater interference when the side and color do not match (♦ ---) than when the color and the side match (---♦). The interesting part is that bilinguals also vary in how well they can perform this task. Those who show more interference are more prone to inadvertently blurt out the name of a picture in the wrong language during language switching tasks. Those who show better performance on control tasks are also better at controlling which language they should produce. These results suggest that individual differences in cognitive control do exist. However, these may not be due to bilingualism per se but may be due to differences in control across any particular group of subjects. Future studies are needed to uncover the conditions under which the bilingual advantage in tasks of cognitive control is or is not observed.

BILINGUAL CONTROL AND LANGUAGE PROFICIENCY

Earlier we pointed out that cognitive control plays an important role in processing a less proficient language. Work in my laboratory has expanded this in two separate studies. Work by Poorna Kushalnagar, a former graduate student of mine, has also investigated the role that language proficiency plays in bimodal bilinguals during attentional tasks (Kushalnagar, Hannay, & Hernandez, 2010). Kushalnagar tested two groups of bilinguals who had high proficiency in both languages and a second group that had high proficiency in only one language. Participants were presented a set of plus signs that occupied both the center and the periphery of the visual field. In one condition, she presented a target that was a different color either in the periphery or the center in separate blocks of trials. In a second condition, participants had stimuli that could appear either in the center or the periphery of the visual field. The results showed that bilinguals with high proficiency in both languages were better at the conditions that required switching between the periphery and the central visual field than those who were less balanced. This suggests that language proficiency may play a role in the types of advantages seen in bimodal bilinguals. Interestingly, it suggests that better bimodal bilingualism may result in differences in how well an individual can track a stimulus across the entire visual field. The surprising part is that this task is visually based. This suggests that the benefits of bimodal bilingualism may be different from those of unimodal bilingualism. It also points to the possibility that proficiency may play a role in the use of cognitive control in bimodal bilinguals.

A second study has found an even more interesting effect of proficiency in a group of child bilinguals. In chapters 3 and 5, I discussed work by Pilar Archila, a graduate student working in my laboratory in collaboration with Jason Zevin (Archila-Suerte et al., 2012). These studies with adults found that AoA and proficiency were reflected in different behavioral and neural indices during phonological processing. Specifically, proficiency was more of a cognitive effect, whereas AoA involved processing the sounds more directly. These types of studies are valuable in that they show how different aspects of development and use influence brain and cognitive systems.

Studies using a retrospective approach in which developmental factors are explored by looking at adult populations with different learning histories are informative but leave out valuable information. Specifically, studies with adults tend to obscure the type of change that occurs across childhood, adolescence, and into adulthood.

To further extend this type of work, Archila asked a group of early Spanish-English second language learners between the ages of 6 and 10 to listen to speech sounds in English while being scanned with fMRI. All of the subjects came from Spanish-speaking homes and had begun to learn English at school during kindergarten at the age of five. For comparison purposes, Archila tested a group of monolingual children across this age range (Archila-Suerte, Zevin, & Hernandez, April 2010). The results for monolingual children confirmed what has been previously found in the literature. Namely, there was increased neural activity in the left and right superior temporal gyrus for speech sounds in both adults and children. There was relatively little change across age for this simple task.

The interesting part of this was in observing how the pattern of neural activity evolved across time in bilinguals. Bilingual children between the ages of six and seven showed similar but weaker neural activity in the superior temporal gyrus compared to that observed in monolingual children. Between the ages of 8 and 10, bilingual children showed a very different pattern of activity. The regions active during speech sound listening included bilateral superior temporal gyrus, the superior and inferior parietal lobule, the inferior frontal gyrus, and the right middle frontal gyrus. As noted above, the superior temporal gyrus is involved in processing speech. However, the additional areas of brain activity are involved in motor planning (inferior frontal gyrus), executive function (middle frontal gyrus), and attention (superior and inferior parietal cortices). These results show that as child second language learners get older and gain experience in a language they switch from the use of first language speech sound representation to a more distributed network of areas. Taken together these results suggest older bilingual children need the support of additional brain areas in order to perceive second language sounds.

As I just noted, the improvement of proficiency that appears in older children involves the recruitment of areas involved in cognitive control. To further identify these effects Archila compared the 8- to 10-year-olds directly with younger children. These comparisons revealed more activity in the left superior temporal gyrus, inferior parietal lobule, dorsolateral prefrontal cortex, and parahippocampal gyrus. We have talked extensively about the inferior parietal lobule and the dorsolateral prefrontal cortex and their involvement in cognitive control. The interesting question is why control is needed. What we observe is that brain areas involved in memory and speech processing are also active. Hence, children are basically forming new speech categories and placing them into memory. Thus the formation of new speech categories during child second language learning naturally leads to the use of cognitive control. Finally, a similar group of early adult bilinguals no longer show the extensive activity seen in the 8- to 10-year-old group. Hence, the use of cognitive control diminishes over time as proficiency in the second language improves.

The results from the studies discussed so far show that the bilingual advantage is a real and significant phenomenon. Bilinguals have to manage two labels for the same thing and appear to handle this transition sooner leading to advantages in early childhood over monolinguals. The advantage also appears to have a possible lower level origin in the fact that bilingual children have to manage two different speech sound systems. The need for control at so many different levels of language may be the reason that such a pervasive use of executive systems in the brain are seen relative to monolinguals especially in childhood.

The findings of control advantages in children and older adults puts to rest any fear that bilingualism and its supposed confusion could ever be seen as a disadvantage. There are still some open questions about the advantage for the Simon task and other interference tasks in the young adult population. However, the presence of this effect in children and older adults is quite clear and merits further study. It is in wading through the fog of this confusion that the bilingual brain learns to handle conflict so well that even nonlinguistic tasks become easier in children and older adults.

BEYOND CONTROL

As I noted earlier, work by Tamar Gollan shows a disadvantage in vocabulary knowledge and use in bilinguals relative to monolinguals. I also discussed the work by Hanako Yoshida and Crystal Tran that shows a clear advantage for bilingual children. Yoshida and Tran took things one step further. Once they had established the fact that there was an advantage, the experimenters wanted to find a possible cause for that advantage.

Yoshida looked at work in the developmental literature. Specifically, she was curious about the fact that control is important in language when having to create a second label for an item. During early development children tend to bias labels toward whole objects (Slobin, 1985). Thus they will tend to think that "dog" refers to the whole dog and not to its tail or its fur. As they get older, children will tend to learn alternative labels for items. For example, they may learn to distinguish between the fluffy dog and the red dog when a parent points "to the fluffy one not the red one." Resolution of this type of competition would benefit from enhanced cognitive control.

Considering this other form of competition, Yoshida and Tran wondered if skill in the hungry fish game would accompany children's ability to learn novel adjectives. To test this latter ability, Yoshida and Tran asked children to look at identical objects that varied in texture or color pattern and that had labels (i.e., this is the firby doggy). They were shown two new objects and instructed to pass them the "firby" one. The interesting part is that bilingual children actually outperformed monolingual children on both the novel adjective and the cognitive control task. In addition, performance on both these tasks was highly correlated in bilinguals but not in monolinguals. In short, it appears that at least during early stage of language learning bilinguals may be used to learning new labels for each item. The flexibility to do this results in the need for further cognitive control. This in turn leads to better performance on tasks that require control such as the hungry fish game.

The ability to learn new vocabulary is not exclusive to bilinguals. A number of studies have shown that vocabulary grows across the life span in monolinguals, with older adults showing higher vocabulary than young adults (Verhaeghen, 2003). Research has also shown that adult second language learners are able to acquire new vocabulary quickly. In a seminal study, Judy McLaughlin, Lee Osterhout, and Albert Kim (2004) recruited three groups of native English speakers having completed 14, 63, or 138 hours of introductory French in college. Participants were shown a set of word pairs comprising words that were related (apple-orange), words that were not related (apple-king), or a word followed by a pseudoword (apple-mave). For native speakers, these type of relationships show a clear signature in the averaged EEG waves. Specifically, the N400 that begins about 200 milliseconds and peaks at 400 milliseconds after presentation of a word is changed by both relatedness and the falseness of the word. Seeing "mave" after "apple" results in the largest N400, "king" after "apple" the second largest, and "orange" after "apple" the smallest. One way to think about this wave is as a marker of relatedness. "Mave" is not related at all to "apple," "king" is somewhat more related since it is actually a word, and "orange" is much more closely related to "apple." Second language learners showed a very similar effect after 138 hours of instruction. That is, in less than one year of introductory French, foreign language learners already know the differences between words and pseudowords. Furthermore, they know a little bit about the meaning of words. This follows nicely from Kroll's idea that second language learners are simply mapping new labels to existing concepts.

The most impressive part, however, was that very early in learning nonnative speaker's brains could already distinguish between words and pseudowords. Why do I say their brains? Because they were incredibly inaccurate at correctly identifying words. But their brain waves showed a difference between these items. In fact, this effect was determined by the number of hours that were spent in instruction. The more hours, the stronger the brain wave difference between pseudowords and real words. Almost immediately, they were starting to form implicit concepts of what a word in French should and should not look like.

Work in the bilingual literature has sought to investigate whether the ability to learn new words might be aided by bilingualism in some form. Work in the 1990s had already found that new foreign language vocabulary learning was aided by having learned several languages in the classroom (Papagno & Vallar, 1995; Van Hell & Mahn, 1997). Work by Cristina Sanz at Georgetown University has found that Catalan-Spanish bilinguals are able to learn English better than Spanish monolinguals (Sanz, 2000).

More recently, work by Viorica Marian and colleagues has found that bilinguals are particularly good at learning new vocabulary. In one study, Kaushanskaya and Marian asked a group of early Spanish-English and Chinese-English bilinguals to learn a set of words that were from a language with different sounding words (Kaushanskaya & Marian, 2009). Participants were taught these words by having them pair the items with the English translation. The results revealed that bilinguals had better memory for newly learned words than monolinguals. A second study conducted by Bartolotti, Marian, and colleagues asked bilinguals and monolinguals to learn two novel languages that were based on Morse code (Bartolotti, Marian, Schroeder, & Shook, 2011). During initial learning there was little interference between the known languages and the novel language, since Morse code is not like any spoken language. In this experiment, bilinguals did better than monolinguals. However, when both groups were taught a new language based on Morse code, then it no longer mattered whether a person was bilingual or monolingual. In this condition, what mattered was whether someone had better cognitive control. Hence, it appears that bilingualism helps with foreign language learning when interference is low. However, under conditions of high interference an individual's ability to exert cognitive control may matter more than whether they are bilingual or not.

Building on the work in novel word learning, two of my graduate students proposed to look at the nature of foreign vocabulary learning in both monolinguals and bilinguals (Bradley, King, & Hernandez, 2013). The specific question we wanted to answer was whether we could identify any difference very early in learning depending on whether a person was bilingual or monolingual. Both monolinguals and bilinguals were asked to study words until they could produce the correct English translation over 90 percent of the time. After reaching 90 percent accuracy, they were placed in the fMRI scanner and asked to decide if the German words or a group of control English words were living or nonliving. The bilingual group, which scored lower in vocabulary in English than the monolingual group, significantly outperformed the monolingual group on the word learning task. They not only learned the words faster but also were faster at identifying them and faster when making the living/nonliving decision. Hence, at least behaviorally, bilinguals may score lower on English vocabulary tests than monolinguals, but are faster at learning new words and are more efficient at processing them.

The efficiency of learning by bilinguals was also observed in the neural activity. Bilinguals showed much smaller differences than monolinguals when comparing German to English words. These differences were also observed when comparing both groups directly. For English words, there was no significant increase in activity when comparing monolinguals to bilinguals or vice-versa. For German words, bilinguals showed increased neural activity in the putamen, an area that is involved in motor control. As we have noted earlier, the putamen is subcortical and is thought to be involved in a more implicit form of motor control. The monolingual group, on the other hand, showed increased activity in the right dorsolateral prefrontal cortex, the anterior cingulate gyrus, and the supplementary motor area. All three of these areas are also known to be involved in the control of language. Unlike the putamen that was observed in bilinguals, the cortical areas observed in monolinguals are known to be involved in a more conscious and explicit form of control.

The results from the neuroimaging study reveal the reason that bilinguals outperform monolinguals. It appears that bilinguals are able to learn new words using more automatic cortical circuitry. This fits in with what Kailyn Bradley observed informally during subject testing. Whereas monolinguals seemed to try to use brute force memory, bilinguals worked to sort the words into different piles. Where monolinguals use conscious control, bilinguals use an implicit system. This finding is interesting because it suggests that the bilingual disadvantage, higher TOT states and lower vocabulary in one language, comes with a huge upside. Bilinguals have more flexible word systems that are exquisitely designed to learn more words. The weaker links are, in fact, more malleable. Hence, the benefit of bilingualism spreads beyond nonlinguistic control tasks. It appears in the learning of novel words.

The results from work in the bilingual literature suggest an interesting link between cognitive control and bilingualism. The bilingual brain works hard to fend off interference between the two languages. This leads to advantages in learning new labels even in a single language. It also leads to impressive gains in the ability to learn a third set of labels. But perhaps the most fascinating aspect is that it also spreads beyond language to control in tasks that have seemingly little to do with language. The enrichment of bilingualism isn't alone. Recent studies have found important cognitive benefits for children who undergo music enrichment programs (Crncec, Wilson, & Prior, 2006). Work in my laboratory finds that college athletes' ability to process sports sounds spreads outside of their domain of expertise. They are able to form visual and auditory images of environmental sounds with significantly less brain activity than that observed in recreational athletes (V. Wagner, Woods, Beilock, & Hernandez, April 2010). But none of these activities involves the range of interaction that is offered when learning a new language. The type of intense adaptation that bilingualism constantly requires is of great benefit to a number of people on the planet. In fact, bilingualism may be the ultimate breakfast of champions.

8

THE FINAL FRONTIER

WE BEGAN THIS journey with three prominent neuropsychologists from the late 19th and early 20th century: Théodule Ribot, Jean-Albert Pitres, and Otto Pötzl. The three ideas they proposed, age of acquisition (AoA), language proficiency, and cognitive control, form the pillars not only of the field but also of this book. In the first section of the book, I discussed how AoA plays an integral part of cognitive processing in both childhood and beyond. What emerges from these chapters is what Ping Li and I have termed the "sensorimotor hypothesis." Things that are learned early in life are more closely tied to the senses and are more likely to require simple motor responses. Things learned later in life are more likely to involve more complex forms of cognitive processing. The divide between early sensorimotor processing and later higher cognitive processing also appears in second language learning. Specifically, researchers find that the processing of speech and even grammar that is based on sounds differs in early and late learners of a second language. Taken together these results suggest that both first and second language acquisition can be viewed as involving a sensorimotor aspect early in life and a more complex form of processing later in life.

In the second section, I built on Pitres's original idea by discussing the role that proficiency or expertise plays in language processing. To expand on this idea I presented the construct of deliberate practice. According to K. Anders Ericsson the most crucial aspect of becoming an expert is that people engage in practice over time with feedback in order to fine-tune their expertise. In the bilingual literature, I discussed how expertise leads to an improvement in language in certain domains such as semantic processing, vocabulary, and certain aspects of grammatical processing. Here I proposed that proficiency or expertise as it is typically discussed in the literature has to do with learning that occurs at levels above basic sensorimotor processing. This also applies to domains traditionally studied in the expertise literature such as chess, music, and sports. Like semantics and more complex grammar, these domains build on smaller building blocks that are acquired early in life. Deliberate practice over time helps to fine-tune these early learned skills to bring about improvement in proficiency.

The final section touches on the notion of cognitive control. Unbeknownst to each another, both Pötzl and Jersild explored the notion of control during the 1920s. Pötzl was seen as a pioneer who, along with Penfield, was the first to conceptualize a neural mechanism for switching between languages. Jersild worked on a parallel track in considering the nature of task switching. A number of studies have established the importance of cognitive control across a wide variety of areas. This includes having to ignore one aspect of a stimulus to attend to another. It also appears in the task switching literature, where having to shift from one rule to another involves both the frontal and parietal lobes. The nature of shifting has also had a profound influence on the bilingual literature, where the notion of language switching has taken on a life of its own. Finally, the need to invoke control appears in conditions of having to make adjustments either to different types of stimuli, whether it be a red light and an incoming phone call, or to different social situations, such as the need to switch to one language with a grandparent and to another with a close friend. Both of these research traditions touch on the importance of adjustments.

AOA, PROFICIENCY, AND CONTROL INTERACT

In this book, I have kept most of my discussion centered on the three key concepts (AoA, proficiency, and control) as mostly separate fields of inquiry. In covering these, the focus has been on each concept separately. However, the truth is that these three aspects of language do not function entirely separately. I have reviewed plenty of cases where these constructs collide in the acquisition of a second language. AoA and proficiency can interact. Individuals who learn a second language earlier in life tend to learn it better than those who start late. However, we saw cases where early language learning coupled with loss led later to better proficiency in a language learned earlier in childhood. Similar examples exist when considering proficiency and control. Control is necessary when processing a less proficient language. Finally, we saw some examples where all three factors interact. One example was in the perception of speech, where early sequential bilinguals showed an increase in control during childhood that then disappeared in adulthood. In this situation, AoA interacts with control and proficiency.

What becomes clear in considering the neural and cognitive bases of bilingualism is that there is a considerable amount of variation in the ways in which two languages can settle into a single individual. In this final chapter, I would like to leave you with a few ideas about how we can conceptualize this complex evolving system. To do so I will introduce two different approaches. One is a more computational approach that seeks to model the way in which two languages are learned and maintained across the life span. The second is a more biological approach that seeks to understand how the brain accomplishes this task.

THE NONLINEAR DYNAMICS OF BILINGUALISM

Thinking of the human mind as a nonlinear dynamical system is nothing new. Work in the late 1980s ushered in a new way of thinking about the mind. In this connectionist or neural network view, processing of information by the human mind occurs via a vast set of interacting systems that are much more interconnected than was previously thought (McClelland & Rumelhart, 1988; Rumelhart & McClelland, 1986). These neural network models do not force structure onto the mind. Rather they constrain the ways in which individual processing units interact and the way that a system takes in information and the types of responses that are generated. Interestingly, these systems will tend to self-organize and "create" structure given the particular output and input functions that are defined. Given the malleability of these models, they are well designed to capture change and have been used to model cognitive development in children.

Thinking of language as the product of nonlinear interactions is nothing new to me. Elizabeth Bates, my doctoral advisor, along with Brain MacWhinney, who is now at Carnegie-Mellon University, proposed a "paleo-connectionist" model, the competition model (Bates & MacWhinney, 1987; Bates & MacWhinney, 1989; Bates et al., 1984; MacWhinney & Bates, 1978; MacWhinney, Plelh, & Bates, 1985). In their view, language development is an interactive process that involves competition between different cues. As they saw it, learning "who did what to whom" emerged through a struggle between different cues to meaning. For example, an English speaker would quickly figure out that the first noun in a sentence is the main actor or subject of the sentence. In Spanish, however, the first noun is not always the actor. Spanish speakers are much more sensitive to what noun matches the verb.

The process of competition also occurs for bilinguals who encounter two different languages. In this case, the different ways of signaling the actor of a sentence may collide. Take, for example, the ungrammatical sentence "The dog are kicking the cows." In this sentence, the verb points to the cows as the agent and the first noun points to the dog as the agent. A monolingual English speaker would choose the dog as the actor most of the time. A Spanish speaker would choose the cows as the actors most of the time. And each group does not understand how the other had such an opposite choice. A bilingual lies somewhere in between (A. E. Hernandez, Bates, & Avila, 1994; Reyes & Hernandez, 2006). He or she would use a combination of these two signals in a way that each monolingual does not. These type of in-between states are very hard to conceptualize within a rigid either/or system. The need for flexibility in the system becomes even more important when thinking about how the use of these signals changes over development.

Connectionist models and their ability to capture the nuances of change present an interesting way to think about bilingualism. This malleability can best be seen in the multiple ways in which a system adapts to learning two languages. More recently researchers have extended the competition model by adopting neural network models. This is captured in recent work by Ping Li, Brian MacWhinney, and colleagues.

In a first simulation, Li, Farkas, and MacWhinney (Li, Farkas, & MacWhinney, 2004) designed a model that was charged with learning vocabulary items in a single language. They did not build any structure into the network. They simply fed the network words that were learned at certain ages. At first the network was shown 50 words that are typically learned by a 2½-year-old. Then it was shown 50 new words typical for a 3-year-old along with the original set of words. These words were from all sorts of classes. There were prepositions, nouns, verbs, and so forth. But there was no information about what kinds of words they were.

Initial words influenced the network strongly, leading to very sudden shifts in organization. Then over time it began to self-organize and form different types of word categories. The network was surprisingly similar in structure to what one would expect. Nouns were closer to each other than verbs, and vice versa. It also showed AoA effects in that early learned words were much more influential than late learned words. The important thing to note is that the network organized on its own through the competitive process of encountering more and more words. There was no need to structure the network in any particular way.

The fact that this self-organizing network could learn a single language vocabulary led Zhao and Li to consider the nature of bilingual language acquisition (Zhao & Li, 2010). To do this they designed a network that mapped word sounds to word meanings. They then introduced two languages to the network either simultaneously or in succession. What became clear was that learning each language at different ages had a very different effect on the ultimate structure of the network. Simultaneous presentation of both languages led to relatively independent maps. Sequential presentation of languages led to the second language being parasitic on the first language.

These networks give us a way to visualize many of the principles that I have discussed in this book. The building up of structure occurs across time and space. As structure begins to emerge each language starts to take up different parts of the network. When learning occurs early, the structure of each set of words can exist relatively independently. When learning occurs later in life, a second language becomes attached to the first language. The second language becomes like a parasite feeding off its host, the first language.

This neural network approach is part of a larger movement that has come to view language and cognition as a dynamic system. The dynamic systems approach is characterized by consideration of how different variables are interconnected. Rather than thinking of language as a single thing, it becomes a combination of things. In this view, there is complete interconnectedness between different variables (Smith & Thelen, 1993, 2003; van Gelder, 1998). Unlike a very discrete system, in which there is a clear relationship between two variables, a dynamic system is difficult to predict because the outcome of one variable can affect the outcome of other variables. These systems are constantly changing. Dynamic systems develop through interactions with their environments and go through constant self-reorganization.

This view describes bilingualism extremely well. As we have seen in this book, there are strong interconnections across multiple systems across time. These interconnections change depending on the particular needs a bilingual has at any particular time. Thus, there may be a stronger reliance on speech sound and motor systems early in life. The sensorimotor aspect of processing in infant bilinguals actually leads to extending the time the brain remains open to recognizing speech sounds from other languages (Petitto et al., 2012). Later frontal systems may be invoked to handle more complex problems but not for speech in a monolingual. However, a bilingual who is learning to digest speech in another language later in life may need to invoke frontal systems initially. What appears in this case is that learning two languages can be seen as a path in which different abilities are combined in different ways at different ages.

FROM COGNITIVE PRINCIPLES TO BRAIN PRINCIPLES

In this book I have also discussed a number of studies that use various forms of brain imaging to investigate the neural bases of cognition, language, and bilingualism. Interestingly, the view from brain science has also begun to consider the dynamic nature of cognition. This can be clearly seen in recent work that looks at individual differences in verbal working memory, the ability to keep language information in mind temporarily. Chantal Prat and colleagues suggest that individual differences show three different neural signatures when comparing more capable and less capable subjects (Prat, 2011). Specifically, those who perform a task better differ in the efficiency, synchronization, and adaptability of the brain regions being used relative to those who do not perform as well. In chapter 4 I discussed an analysis by Prat and colleagues in which those with higher vocabularies show less neural activity in brain areas in the left frontal and parietal lobe when performing reading comprehension tasks (Prat & Just, 2011). Hence, the brains of experts are more efficient in that there is relatively less activity compared to that seen in those with less language expertise. Similar reductions in neural activity have been found in studies that look at brain activity during performance of standard intelligence tests (Neubauer & Fink, 2009).

One of the main problems with comparisons of neural efficiency is that they can only be conducted when performance in two groups is equated. When tasks are complex, neural activity in a participant may need to call on a brain area or particular network of areas in order to accomplish the task successfully. The change that occurs when tasks become more difficult has been termed "adaptability" (Schafer, 1982). The nature of adaptability can be seen in a number of studies. For example, Marcel Just from Carnegie-Mellon University, along with a number of his colleagues, found that increasing the difficulty of a grammatical task led to stronger activity in right Broca's area (Just, Carpenter, Keller, Eddy, & Thulborn, 1996). Work along these lines was also recently conducted by Prat along with Marcel Just (Prat & Just, 2011). In this study, participants were asked to perform a working memory test and then separated on their ability to store information temporarily. Later these participants were asked to read sentences with simple or complex grammar. Interestingly, the participants with higher working memory capacity were able to perform their tasks better and showed increased activity in control areas in the frontal cortex. Those with lower working memory did not perform as well and showed less activity in the frontal cortex and other control related areas. In short, those with better adaptability appear to show changes in neural activity in order to perform a task more successfully. Furthermore, those with more capacity also exhibit more adaptability. In short, capacity and adaptability appear to go hand in hand.

The third and final principle that Prat uses to describe individual differences has to do with the synchronization of brain areas. When comparing those with high and low capacity, Prat found that the former had a stronger coupling of activity in left Broca's and Wernicke's area relative to those with low capacity. This tight coupling remained even when an additional distracting task was introduced or the sentences were more difficult to process. This suggests that those with higher capacity have tighter coupling between areas.

The interesting part of this framework is that it makes no direct references to particular brain areas. The notions of efficiency, adaptability, and synchronization can be applied to a number of different types of skills. Each of these skills could involve a number of different brain networks. Hence, rather than thinking about the brain as some form of mental storage facility, we begin to think about it as a set of networks that can be applied to different skills. The notions of efficiency, adaptability, and synchronization offer us a way to think about how two languages are coded in one brain within a more general framework. The problem with bilinguals is that there is considerable change that happens over time. This is where we can consider the notion of adaptability within a broader context.

Finally, the work by Prat, Just, and colleagues also reveals an interesting twist on the instantiation of skills in the brain by emphasizing the dynamic aspects of processing. Rather than skilled and unskilled individuals only showing simple differences in efficiency, there appear to be clear differences in the way that this efficiency plays out as tasks become more complex. Those with better skills are able to adjust or adapt to circumstances better. This dynamic interplay between an individual and the challenges of the environment provides an interesting way to think about bilingualism, the brain, and the nature of skilled performance.

METAPHORS FOR THE MIND

So far I have proposed two frameworks that consider the nature of change and adaptability. You might be wondering why I might be leaning on network modeling and dynamic systems theory in this final section of the book. Neural network modeling as an extension of dynamic systems theory has had another effect on the nature of language processing. It has begun to change the very language that is used to describe cognitive processes. This includes words like "noise," "competition," "entrenchment," "parasitism," and "adaptability." These words are not ones we typically associate with computers. Rather they are terms that are reserved for the description of biological entities. Natural systems tend to be noisy. There is competition between species. Entrenchment might describe what happens when a boat gets caught in a vortex and is unable to escape. The use of a dynamic systems approach equates bilingualism with a natural system (de Bot, Lowie, & Verspoor, 2007).

The field of cognitive psychology took an interesting turn when it began to morph into two separate fields. First, cognitive science used the power of computers to simulate the human mind. But in using computers, they abandoned the view that the mind was like a computer. Rather these computer models began to look more like natural systems. Second, cognitive neuroscience began to search for the brain bases of mental processing. By using magnets, electrodes, and optical sensors we are able today to see the intricacies of the human brain in action.

Over the years there has been strong resistance to a purely computational view of the mind by a group of neo-Piagetians who view development in a very different manner. Among these was Elizabeth Bates, my doctoral advisor, who devoted her career to championing the idea that language was a "new machine built out of old parts." Language development rides along with all sorts of other cognitive milestones that are not directly related to language development. Jeff Elman and a group of colleagues began to have more and more discussions about this topic. The culmination of this effort can be seen in the 1996 book *Rethinking Inateness* (Elman, Bates, Johnson, & Karmiloff-Smith, 1996). In this volume, the authors propose a very ambitious alternative view in which biology and environment perform an intricate waltz across time on their way to the eventual destination of adulthood.

A nice example of the type of complex interactions that take place across development can be seen in the domain of face perception. As I noted in earlier chapters, newborns are able to track faces within minutes of birth (J. Morton & Johnson, 1991). This supports the notion of an innate brain mechanism that is involved in face perception. There is simply no exposure to faces in the womb, and there are no mirrors, so newborns cannot even see their own faces. The fact that newborns look at faces longer does not mean they have all of the aspects of face recognition worked out. Newborns do not have adult-like face recognition. In fact, they are able to see the outline of a face in a very schematic manner from a very short distance. Over development children acquire considerable expertise in face recognition. But this process happens slowly over many years. Even in middle childhood there is still refinement in face recognition ability (Mondloch et al., 2010). Behaviorally, 10-year-old children still differ from adults. Some studies even report a drop in face recognition ability in puberty relative to younger children (Carey, Diamond, & Woods, 1980; McGivern, Andersen, Byrd, Mutter, & Reilly, 2002).

John Morton and Mark Johnson, one of the authors of *Rethinking Innateness*, have proposed that face recognition comes to be modulated by very different systems (J. Morton & Johnson, 1991) that are involved in early face detection, with later systems being involved in learning about faces. In 2005 Johnson further expanded this view by suggesting that infants made use of a subcortical route (M. H. Johnson, 2005). Interestingly, there is evidence that adults can use this route to recognize stimuli that may not be available to conscious thought. The use of this subcortical route gets infants going in the right direction. Over time other cortical areas are recruited for face processing. These areas are then slowly integrated into a network that handles the processing of faces (Cohen Kadosh, Cohen Kadosh, Dick, & Johnson, 2011). Notice, however, that we are taking about a core system that gets built up over time. Nowhere does Johnson claim that newborns have knowledge about faces. Rather they have a set of initial biases that make them attend to faces much longer than they attend to other types of objects in the environment. Language shows a similar type of developmental trajectory. Early learning is based on very early sensorimotor systems that get built up over time. As with face processing there is evidence that the processing of speech sounds continues to improve across childhood. Furthermore, it appears to be open even in some adults, who can learn a second language with near perfect proficiency (Abrahamsson & Hyltenstam, 2008; Hyltenstam, Abrahamsson, Doughty, & Long, 2003).

THE BILINGUAL BRAIN

In the previous section, I showed how a system can be built up over time. What becomes clear is that AoA, proficiency, and cognitive control are interconnected in the bilingual brain. The problem is how to visualize this process. What metaphor might help to elucidate the types of changes that occur in the neurons of bilinguals as they learn two languages? Struggling with this question, I tried to create a visual image of what I was talking about. To do this I began to visualize the different stages of cognitive development as a set of concentric circles.

Each of these circles is designed to represent one of Piaget's stages of development. At the core of processing is sensorimotor processing. This involves the types of loops that are used for the basic functions that I talk about in chapters 2 and 3. Things like the perception and production of speech sounds in early learners of a language. The next circle represents precognitive processing. This involves slightly more complex skills such as the processing of sentences or determining the past tense of an irregular verb. Finally, we have cognitive processing, which involves a more abstract level that

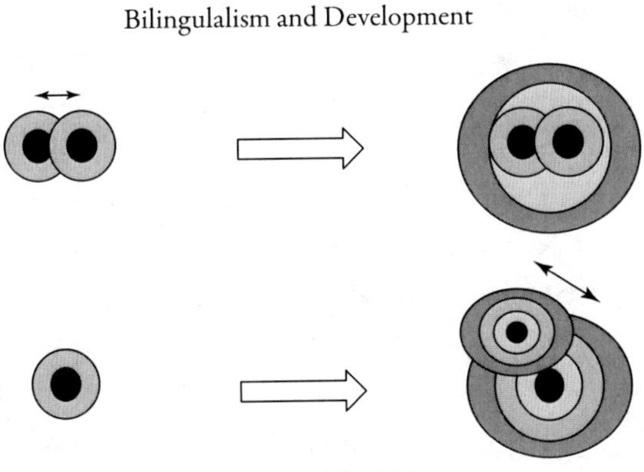

FIGURE 8.1 Representation of a cellular-like model of bilingualism.

relies on complex loops with the need of cognitive control mechanisms to sort it out. This concentric model is a very rough approximation of what is happening to an individual across development. It is not a precise demonstration of the complexity I have described in this book. It is just a way to grasp it more concretely.

An interesting thing happened when I began to think about how to visualize bilingualism (Figure 8.1). For early bilinguals we see what looks like a fused core with the later layers of cognition built around it. These two languages are distinct but they share a set of core processes. Late bilinguals, on the other hand, may have a much less developed core in their second language. By the time second language learning begins, a single system exists with considerable structure already in place. Exposure to a second language is a form of collision in which the new sphere attaches itself to an already existing system. What results is a form of parasitism in which the "invading" language carves out space in the existing system.

When I first explained this metaphor to Lynette Austin a few years ago, I thought it looked like a planet. Early bilingualism would be like two twin planets growing together. Late bilingualism would be like a large asteroid colliding into an existing planet. The planetary metaphor was a nice one, but it seemed incomplete. It captured the natural evolving nature of language but it was a bit brittle. Planets change but they don't fuse together. If two planets collided you would get considerable destruction and there might be an asteroid belt left. The image of learning a second language late in life leading to space dust was not really a useful thing in my thinking. As I stared at the concentric circles it began to occur to me that they resembled something else. When you look at them carefully, these concentric circles look like growing collections of cells. The collisions were not between planets but between evolving entities.

That is when I realized that the best metaphor for the bilingual brain and mind was biological. With this cellular metaphor in hand, I began to think about how exposure to language is so similar to what a cell or any growing organism might go through. This helped in some ways to explain some of the paradoxes of bilingualism. One possible view of the AoA literature is that early exposure leads to better learning of a language. The problem is that this is only true with continued exposure. Early learners can pick up the accent of a language. However, that might be the only thing they learn. What is left is a huge set of concentric circles with a very small circle in the core. This small circle would be the second language. A small orphan phonology stuck in between the extensive network that constitutes the person's dominant language. Continued exposure also played a role in research spearheaded by Pallier with Korean adoptees in France. In this case, the original structure of language was simply adapted to the new circumstances. This type of complete transformation is indeed surprising. The complex interactions in language learning are also seen in adults who seemingly lose access to a language when placed in extreme immersion.

Rather than thinking of bilingualism and second language learning as something abstract and computer-like, I propose that we finally accept that it is biological. It is the product of neurons that fire in a vast interconnected network. Each language is like a living being inside of us. Some parts of the network are unique. Not in a content-addressable kind of way. There is no English, Russian, or Mandarin space in the brain. Rather there are English, Russian, and Mandarin states. These unique states are spatial and temporal in nature. They are like schools of fish that travel in waves across the brain. In this book, I have provided a landscape for the intricacy needed to handle two languages in one brain. The most exciting part is that our field is really in its infancy. The tools that are just on the horizon will bring together the biological brain signals with computational network models. I imagine a not too distant future where individuals will be able to see visual representation of their brain's language networks. In that moment, we'll catch a glimpse of perhaps our biggest achievement as a species: the ability to cram not one but two languages (or more!) in one brain.

Bibliography

Abrahamsson, N., & Hyltenstam, K. (2008). The robustness of aptitude effects in near-native second language acquisition. *Studies in Second Language Acquisition, 30*, 481–509.

Abutalebi, J. (2008). Neural aspects of second language representation and language control. *Acta Psychologica (Amsterdam), 128*, 466–478.

Abutalebi, J., Cappa, S. F., & Perani, D. (2001). The bilingual brain as revealed by functional neuroimaging. *Bilingualism: Language and Cognition, 4*, 179–190.

Abutalebi, J., Della Rosa, P. A., Tettamanti, M., Green, D. W., & Cappa, S. F. (2009). Bilingual aphasia and language control: A follow-up fMRI and intrinsic connectivity study. *Brain and Language, 109*, 141–156.

Abutalebi, J., & Green, D. (2007). Bilingual language production: The neurocognition of language representation and control. *Journal of Neurolinguistics, 20*, 242–275.

Abutalebi, J., & Green, D. W. (2008). Control mechanisms in bilingual language production: Neural evidence from language switching studies. *Language and Cognitive Processes, 23*, 557–582.

Allport, D. A., Styles, E. A., & Hsieh, S. (1994). Shifting intentional set: Exploring the dynamic control of tasks. In M. M. E. Carlo Umilta (Ed.), *Attention and performance 15: Conscious and nonconscious information processing. Attention and performance series* (pp. 421–452). Cambridge, MA: MIT Press.

Altarriba, J. (1992). The representation of translation equivalents in bilingual memory. In R. J. Harris (Ed.), *Cognitive processing in bilinguals* (Vol. 83, pp. 157–174). Amsterdam: North-Holland.

Archila-Suerte, P., Zevin, J., Bunta, F., & Hernandez, A. E. (2012). Age of acquisition and proficiency in a second language independently influence the perception of non-native speech. *Bilingualism: Language and Cognition, 15*, 190–201.

Archila-Suerte, P., Zevin, J., & Hernandez, A. E. (April 2010). *Developmental effects in the speech perception of bilingual children.* Paper presented at the Annual Meeting of the Cognitive Neuroscience Society, San Francisco.

Ashby, F. G., & Crossley, M. J. (2012). Automaticity and multiple memory systems. *Wiley Interdisciplinary Reviews: Cognitive Science, 3*, 363–376.

Asher, J. J., & Garcia, R. (1969). The optimal age to learn a foreign language. *Modern Language Journal, 53*, 334–341.

Au, R., Joung, P., Nicholas, M., & Obler, L. K. (1995). Naming ability across the adult life span. *Aging and Cognition, 2*, 300–311.

Au, T. K.-F., Knightly, L. M., Jun, S.-A., & Oh, J. S. (2002). Overhearing a language during childhood. *Psychological Science, 13*, 238–243.

Austin, D. L. B. (2007). *The impact of age of acquisition of English, extent of language use, and first language proficiency on the bilingual language skills of Spanish-English bilinguals.* Ph.D. dissertation. University of Houston. Houston, TX. 2007.

Badre, D., Poldrack, R. A., Pare-Blagoev, E. J., Insler, R. Z., & Wagner, A. D. (2005). Dissociable controlled retrieval and generalized selection mechanisms in ventrolateral prefrontal cortex. *Neuron, 47*, 907–918.

Baillargeon, R. E. (2004). Infants' reasoning about hidden objects: Evidence for event-general and event-specific expectations. *Developmental Science, 7*, 391–424.

Banks, M. S., Aslin, R. N., & Letson, R. D. (1975). Sensitive period for the development of human binocular vision. *Science, 190*, 675–677.

Bartolotti, J., Marian, V., Schroeder, S. R., & Shook, A. (2011). Bilingualism and inhibitory control influence statistical learning of novel word forms. *Frontiers in Psychology, 2*, 324.

Bates, E. (1974). The acquisition of conditional verbs by Italian children. *Proceedings from the 10th Regional Meetings of the Chicago Linguistic Society* (pp. 27–36). Chicago: University of Chicago.

Bates, E., Benigni, L., Bretherton, I., Camaioni, L., & Volterra, V. (1979). *The emergence of symbols: Cognition and communication in infancy.* New York: Academic Press.

Bates, E., Bretherton, I., & Snyder, L. (1988). *From first words to grammar: Individual differences and dissociable mechanisms.* New York: Cambridge University Press.

Bates, E., D'Amico, S., Jacobsen, T., Szekely, A., Andonova, E., Devescovi, A., . . . Tzeng, O. (2003). Timed picture naming in seven languages. *Psychonomic Bulletin and Review, 10*, 344–380.

Bates, E., & MacWhinney, B. (1987). Competition, variation and language learning. In B. MacWhinney (Ed.), *Mechanisms of language acquisition* (pp. 157–193). Hillsdale, NJ: Erlbaum.

Bates, E., & MacWhinney, B. (1989). *The crosslinguistic study of sentence processing.* New York: Cambridge University Press.

Bates, E., MacWhinney, B., Caselli, C. M., Devescovi, A., Natale, F., & Venza, V. (1984). A cross-linguistic study of the development of sentence interpretation strategies. *Child Development, 55*, 341–354.

Bates, E., Wilson, S. M., Saygin, A. P., Dick, F., Sereno, M. I., Knight, R. T., & Dronkers, N. F. (2003). Voxel-based lesion-symptom mapping. *Nature Neuroscience, 6*, 448–450.

Bavin, E. L. (1992). The acquisition of Walpiri. In D. I. Slobin (Ed.), *The crosslinguistic study of language acquisition* (Vol. 3, pp. 309–371). Hillsdale, NJ: Erlbaum.

BBC (Producer). (2010). Case Study: HM—The Man Who Couldn't Remember. *Mind Changers.* Retrieved from http://www.bbc.co.uk/programmes/b00t6zqv

Bechara, A., Tranel, D., & Damasio, H. (2000). Characterization of the decision-making deficit of patients with ventromedial prefrontal cortex lesions. *Brain, 123* (Pt 11), 2189–2202.

Beilock, S. L., Carr, T. H., MacMahon, C., & Starkes, J. L. (2002). When paying attention becomes counterproductive: Impact of divided versus skill-focused attention on novice and experienced performance of sensorimotor skills. *Journal of Experimental Psychology: Applied, 8*, 6–16.

Benasich, A. A., Gou, Z., Choudhury, N., & Harris, K. D. (2008). Early cognitive and language skills are linked to resting frontal gamma power across the first 3 years. *Behavioural Brain Research, 195*, 215–222.

Besner, D., & Smith, M. (1992). Models of visual word recognition: When obscuring the stimulus yields a clearer view. *Journal of Experimental Psychology: Learning, Memory, and Cognition, 18*, 468–482.

Best, C. T. (1988). The emergence of cerebral asymmetries in early human development: A literature review and a neuroembryological model. In D. L. Molfese & S. J. Segalowitz (Eds.), *Brain lateralization in children: Developmental implications* (pp. 5–34). New York: Guilford.

Bialystok, E. (1988). Levels of bilingualism and levels of linguistic awareness. *Developmental Psychology, 24*, 560–567.

Bialystok, E. (1999). Cognitive complexity and attentional control in the bilingual mind. *Child Development, 70*, 636–644.

Bialystok, E. (2007). Cognitive effects of bilingualism: How linguistic experience leads to cognitive change. *International Journal of Bilingual Education and Bilingualism, 10*, 210–223.

Bialystok, E., Craik, F. I. M., & Freedman, M. (2007). Bilingualism as a protection against the onset of symptoms of dementia. *Neuropsychologia, 45*, 459–464.

Bialystok, E., Craik, F. I. M., Klein, R., & Viswanathan, M. (2004). Bilingualism, aging, and cognitive control: Evidence from the Simon task. *Psychology and Aging, 19*, 290–303.

Bialystok, E., Craik, F. I. M., & Ryan, J. (2006). Executive control in a modified antisaccade task: Effects of aging and bilingualism. *Journal of Experimental Psychology: Learning, Memory, and Cognition, 32*, 1341–1354.

Bialystok, E., & Hakuta, K. (1994). *In other words: The science and psychology of second-language acquisition*. New York: Basic Books.

Bialystok, E., & Harris, R. J. (1992). Selective attention in cognitive processing: The bilingual edge. In R. J. Harris (Ed.), *Cognitive processing in bilinguals* (pp. 501–513). Amsterdam: North-Holland.

Bialystok, E., & Martin, M. M. (2004). Attention and inhibition in bilingual children: Evidence from the dimensional change card sort task. *Developmental Science, 7*, 325–339.

Bianchi, L. (1886). Un caso di sordita verbale. *Rivista sperimentale di Freniatria, 12*, 57–71.

Birdsong, D. (2006). Age and second language acquisition and processing: A selective overview. *Language Learning, 56*, 9–49.

Birdsong, D., & Molis, M. (2001). On the evidence for maturational constraints in second-language acquisition. *Journal of Memory and Language, 44*, 235–249.

Bjork, R. A., & Benjamin, A. S. (2011). On the symbiosis of remembering, forgetting, and learning. In A. S. Benjamin (Ed.), *Successful remembering and successful forgetting: A festschrift in honor of Robert A. Bjork*. (pp. 1–22). New York: Psychology Press.

Bookheimer, S. (2002). Functional MRI of language: new approaches to understanding the cortical organization of semantic processing. *Annual Review of Neuroscience, 25*, 151–188.

Boroditsky, L., Schmidt, L. A., & Phillips, W. (2003). Sex, syntax and semantics. In D. Gentner & S. Goldin-Meadow (Eds.), *Language in mind: Advances in the study of language and thought.* (pp. 61–79). Cambridge, MA: MIT Press.

Bosch, L., & Sebastian-Galles, N. (2001). Evidence of early language discrimination abilities in infants from bilingual environments. *Infancy, 2*, 29–49.

Boudreault, P., & Mayberry, R. (2006). Grammatical processing in American Sign Language: Effects of age of first language acquisition and syntactic complexity. *Journal of Language and Cognitive Processes, 21*, 608–635.

Bradley, K. A., King, K. E., & Hernandez, A. E. (2013). Language experience differentiates prefrontal and subcortical activation of the cognitive control network in novel word learning. [Research Support, N.I.H., Extramural]. *Neuroimage, 67*, 101–110. doi: 10.1016/j.neuroimage.2012.11.018

Brainard, M., & Knudsen, E. (1998). Sensitive periods for visual calibration of the auditory space map in the barn own optic tectum. *Journal of Neuroscience, 18*, 3929–3942.

Brainard, M. S., & Doupe, A. J. (2002). What songbirds teach us about learning. *Nature, 417*, 351–358.

Broca, P. (1988). On the speech center. In L. T. Benjamin (Ed.), *A history of psychology: Original sources and contemporary research*. New York: McGraw-Hill.

Brown, P. (2001). Learning to talk about motion UP and DOWN in Tzeltal: Is there a language-specific bias for verb learning? In M. Bowerman & S. C. Levinson (Eds.), *Language acquisition and conceptual development* (pp. 512–543). Cambridge: Cambridge University Press.

Cabeza, R., Daselaar, S. M., Dolcos, F., Prince, S. E., Budde, M., & Nyberg, L. (2004). Task-independent and task-specific age effects on brain activity during working memory, visual attention and episodic retrieval. *Cerebral Cortex, 14*, 364–375.

Camerer, C. F., Johnson, E. J., Ericsson, K. A., & Smith, J. (1991). The process-performance paradox in expert judgment: How can experts know so much and predict so badly? *Toward a general theory of expertise: Prospects and limits.* (pp. 195–217). New York: Cambridge University Press.

Campbell, S., & Whitaker, H.A. (1986). Cortical maturation and developmental neurolinguistics. In J. E. Obrzut & G. W. Hynd (Eds.), *Child neuropsychology: Vol. I. Theory and research* (pp. 55–72). New York: Academic Press.

Candland, D. K. (1993). *Feral children and clever animals: Reflections on human nature.* New York: Oxford University Press.

Carey, S., Diamond, R., & Woods, B. (1980). Development of face recognition: A maturational component? *Developmental Psychology, 16*, 257–269.

Carpenter, P. A., Just, M. A., Keller, T. A., Eddy, W., & Thulborn, K. (1999). Graded functional activation in the visuospatial system with the amount of task demand. *Journal of Cognitive Neuroscience, 11*, 9–24.

Chase, W. G., & Simon, H. A. (1973a). The mind's eye in chess. In W. G. Chase (Ed.), *Visual information processing* (pp. 215–281). New York: Academic Press.

Chase, W. G., & Simon, H. A. (1973b). Perception in chess. *Cognitive Psychology, 4*, 55–81.

Chauncey, K., Midgley, K. J., Grainger, J., van Heuven, W., & Holcomb, P. J. (2004). Cross-language neighborhood effects in bilinguals: An electrophysiological investigation. Cognitive Neuroscience Society 2004 annual meeting program [Supplement]. *Journal of Cognitive Neuroscience.*

Chee, M. W., Hon, N., Lee, H. L., & Soon, C. S. (2001). Relative language proficiency modulates BOLD signal change when bilinguals perform semantic judgments. *NeuroImage, 13*, 1155–1163.

Chee, M. W., Soon, C. S., Lee, H. L., & Pallier, C. (2004). Left insula activation: a marker for language attainment in bilinguals. *Proceedings of the National Academy of Sciences of the United States of America, 101*, 15265–15270.

Cohen Kadosh, K., Cohen Kadosh, R., Dick, F., & Johnson, M. H. (2011). Developmental changes in effective connectivity in the emerging core face network. *Cerebral Cortex, 21*, 1389–1394.

Costa, A., Santesteban, M., & Ivanova, I. (2006). How do highly proficient bilinguals control their lexicalization process? Inhibitory and language-specific selection mechanisms are both functional. *Journal of Experimental Psychology: Learning, Memory, and Cognition, 32*, 1057–1074.

Crinion, J., Turner, R., Grogan, A., Hanakawa, T., Noppeney, U., Devlin, J. T.,... Price, C. J. (2006). Language control in the bilingual brain. *Science, 312*, 1537–1540.

Crncec, R., Wilson, S. J., & Prior, M. (2006). The cognitive and academic benefits of music to children: Facts and fiction. *Educational Psychology, 26*, 579–594.

Crone, E. A., Bunge, S. A., van der Molen, M. W., & Ridderinkhof, K. R. (2006). Switching between tasks and responses: A developmental study. *Developmental Science, 9*, 278–287.

Crone, E. A., Donohue, S. E., Honomichl, R., Wendelken, C., & Bunge, S. A. (2006). Brain regions mediating flexible rule use during development. *Journal of Neuroscience, 26*, 11239–11247.

Curtiss, S., Fromkin, V., Krashen, S., Rigler, D., & Rigler, M. (1974). The linguistic development of Genie. *Language, 50*, 528–554.

Dalrymple-Alford, E. C., & Aamiry, A. (1969). Language and category clustering in bilingual free recall. *Journal of Verbal Learning and Verbal Behavior, 8*, 762–768.

Daneman, M. (1982). The measurement of reading comprehension: How not to trade construct validity for predictive power. *Intelligence, 6*, 331–345.

Daneman, M. (1991). Working memory as a predictor of verbal fluency. *Journal of Psycholinguistic Research, 20*, 445–464.

Daneman, M., & Carpenter, P. A. (1980). Individual differences in working memory and reading. *Journal of Verbal Learning and Verbal Behavior, 19*, 450–466.

Daneman, M., & Carpenter, P. A. (1983). Individual differences in integrating information between and within sentences. *Journal of Experimental Psychology: Learning, Memory, and Cognition, 9*, 561–584.

Daneman, M., & Green, I. (1986). Individual differences in comprehending and producing words in context. *Journal of Memory and Language, 25*, 1–18.

Daneman, M., & Merikle, P. M. (1996). Working memory and language comprehension: A meta-analysis. *Psychonomic Bulletin and Review, 3*, 422–433.

Daneman, M., Nemeth, S., Stainton, M., & Huelsmann, K. (1995). Working memory as a predictor of reading achievement in orally educated hearing-impaired children. *Volta Review, 97*, 225–241.

Davidson, M. C., Amso, D., Anderson, L. C., & Diamond, A. (2006). Development of cognitive control and executive functions from 4 to 13 years: Evidence from manipulations of memory, inhibition, and task switching. *Neuropsychologia, 44*, 2037–2078.

Dawes, R. M. (1996). *House of cards: Psychology and psychotherapy built on myth*. New York: Free Press.

de Bot, K., Lowie, W., & Verspoor, M. (2007). A dynamic systems theory approach to second language acquisition. *Bilingualism: Language and Cognition, 10*, 7–21.

de Groot, A. (1978). *Thought and Choice in Chess*. The Hague: Mouton.

Demuth, K., Patrolia, M., Song, J. Y., & Masapollo, M. (2012). The development of articles in children's early Spanish: Prosodic interactions between lexical and grammatical form. *First Language, 32*, 17–37. doi: 10.1177/0142723710396796

Deutsch, D., Henthorn, T., Marvin, E., & Xu, H.-S. (2006). Absolute pitch among American and Chinese conservatory students: Prevalence differences, and evidence for a speech-related critical period. *Journal of the Acoustical Society of America, 119*, 719–722.

Devlin, J. T., Jamison, H. L., Gonnerman, L. M., & Matthews, P. M. (2006). The role of the posterior fusiform gyrus in reading. *Journal of Cognitive Neuroscience, 18*, 911–922.

Doll, J. R., & Mayr, U. (1987). Intelligenz und Schachleistung—eine Untersuchung an Schachexperten. *Psychologische Beitrage, 29*, 270–289.

Dowens, M. G., Vergara, M., Barber, H. A., & Carreiras, M. (2010). Morphosyntactic processing in late second-language learners. *Journal of Cognitive Neuroscience, 22*, 1870–1887.

Drews, F. A., Yazdani, H., Godfrey, C. N., Cooper, J. M., & Strayer, D. L. (2009). Text messaging during simulated driving. *Human Factors, 51*, 762–770.

Dronkers, N. F. (1996). A new brain region for coordinating speech articulation. *Nature, 384*, 159–161.

Eimas, P. D., Siqueland, E. R., Jusczyk, P., & Vigorito, J. (1971). Speech perception in infants. *Science, 171*, 303–306.

Elbert, T., Pantev, C., Wienbruch, C., Rockstroh, B., & Taub, E. (1995). Increased cortical representation of the fingers of the left hand in string players. *Science, 270*, 305–307.

Ellis, A. W., & Lambon Ralph, M. A. (2000). Age of acquisition effects in adult lexical processing reflect loss of plasticity in maturing systems: Insights from connectionist networks. *Journal of Experimental Psychology: Learning, Memory, and Cognition, 26*, 1103–1123.

Ellis, H. D., & Florence, M. (1990). Bodamer's (1947) paper on prosopagnosia. *Cognitive Neuropsychology, 7*, 81–105.

Elman, J. L., Bates, E. A., Johnson, M. H., & Karmiloff-Smith, A. (1996). *Rethinking innateness: A connectionist perspective on development*: Cambridge, MA: MIT Press.

Elston-Guettler, K. E., Paulmann, S., & Kotz, S. A. (2005). Who's in control? proficiency and L1 influence on L2 processing. *Journal of Cognitive Neuroscience, 17*, 1593–1610.

Emmorey, K., Luk, G., Pyers, J. E., & Bialystok, E. (2008). The source of enhanced cognitive control in bilinguals: evidence from bimodal bilinguals. *Psychological Science, 19*, 1201–1206.

Ericsson, K. A., Charness, N., Feltovich, P. J., & Hoffman, R. R. (2006). *The Cambridge handbook of expertise and expert performance*. New York: Cambridge University Press.

Ericsson, K. A., Krampe, R. T., & Tesch-Römer, C. (1993). The role of deliberate practice in the acquisition of expert performance. *Psychological Review, 100*, 363–406.

Ericsson, K. A., & Simon, H. A. (1980). Verbal reports as data. *Psychological Review, 87*, 215–251.

Ericsson, K. A., & Simon, H. A. (1993). *Protocol analysis: Verbal reports as data* (rev. ed.). Cambridge, MA US: The MIT Press.

Fabbro, F., Skrap, M., & Aglioti, S. (2000). Pathological switching between languages after frontal lesions in a bilingual patient. *Journal of Neurology, Neurosurgery and Psychiatry, 68*, 650–652.

Fagiolini, M., Pizzorusso, T., Berardi, N., Domenici, L., & Maffei, L. (1994). Functional postnatal development of the rat primary visual cortex and the role of visual experience: dark rearing and monocular deprivation. *Vision Research, 34*, 709–720.

Farah, M. J. (2004). *Visual agnosia.* Cambridge, Mass.: MIT Press.

Ferlazzo, F., Fagioli, S., Di Nocera, F., & Sdoia, S. (2008). Shifting attention across near and far spaces: Implications for the use of hands-free cell phones while driving. *Accident Analysis and Prevention, 40*, 1859–1864.

Fernandes, M. A., Craik, F., Bialystok, E., & Kreuger, S. (2007). Effects of bilingualism, aging, and semantic relatedness on memory under divided attention. *Canadian Journal of Experimental Psychology/Revue canadienne de psychologie expérimentale, 61*, 128–141.

Festman, J., & Munte, T. F. (2012). Cognitive control in Russian-German bilinguals. *Frontiers in Psychology, 3*, 115.

Festman, J., Rodriguez-Fornells, A., & Munte, T. F. (2010). Individual differences in control of language interference in late bilinguals are mainly related to general executive abilities. *Behavioral and Brain Functions, 6*, 5.

Fiebach, C. J., Friederici, A. D., Muller, K., von Cramon, D. Y., & Hernandez, A. E. (2003). Distinct brain representations for early and late learned words. *NeuroImage, 19*, 1627–1637.

Finn, A. S., Sheridan, M. A., Kam, C. L., Hinshaw, S., & D'Esposito, M. (2010). Longitudinal evidence for functional specialization of the neural circuit supporting working memory in the human brain. *Journal of Neuroscience, 30*, 11062–11067.

Fitts, P. M., & Posner, M. I. (1967). Learning and skilled performance in human performance. *Belmont, CA: Brock-Cole.*

Flege, J. E. & MacKay, I. R, A. (2004). Perceiving vowels in a second language. *Studies in Second Language Acquisition, 26*, 1–34.

Foss, D., & Speer, S. R. (1983). Global and local context effects in sentence processing. In R. H. D. Palermo (Ed.), *Cognition and the symbolic processes: Applied ecological perspectives.* Hillsdale, NJ: Erlbaum

French, R. M., & Jacquet, M. (2004). Understanding bilingual memory: Models and data. *Trends in Cognitive Sciences, 8*, 87–93.

Friederici, A. D. (2002). Towards a neural basis of auditory sentence processing. *Trends in Cognitive Sciences, 6*, 78–84.

Friederici, A. D. (2006). What's in control of language? *Nature Neuroscience, 9*, 991–992.

Friederici, A. D., Hahne, A., & Mecklinger, A. (1996). Temporal structure of syntactic parsing: early and late event-related brain potential effects. *Journal of Experimental Psychology: Learning, Memory, and Cognition, 22*, 1219–1248.

Friederici, A. D., & Meyer, M. (2004). The brain knows the difference: Two types of grammatical violations. *Brain Research, 1000*, 72–77.

Garbin, G., Costa, A., Sanjuan, A., Forn, C., Rodriguez-Pujadas, A., Ventura, N.,... Åvila, C. (2011). Neural bases of language switching in high and early proficient bilinguals. *Brain and Language, 119*, 129–135.

Gare, S. (2009). Success is all in the mind. *The Weekend Australian,* January 24–25, p. 1.

Gauthier, I., & Nelson, C. A. (2001). The development of face expertise. *Current Opinion in Neurobiology, 11*, 219–224.

Gauthier, I., Skudlarski, P., Gore, J. C., & Anderson, A. W. (2000). Expertise for cars and birds recruits brain areas involved in face recognition. *Nature Neuroscience, 3*, 191–197.

Gauthier, I., & Tarr, M. J. (2002). Unraveling mechanisms for expert object recognition: bridging brain activity and behavior. *Journal of Experimental Psychology: Human Perception and Performance, 28*, 431–446.

Gazzaniga, M. S., Ivry, R. B., & Mangun, G. R. (2009). *Cognitive neuroscience: the biology of the mind*. New York: Norton.

Genesee, F. (1989). Early bilingual development: One language or two? *Journal of Child Language, 16*, 161–179.

Gerken, L. (1994). A metrical template account of children's weak syllable omissions from multisyllabic words. *Journal of Child Language, 21*, 565–584.

Gerken, L., & McIntosh, B. J. (1993). Interplay of function morphemes and prosody in early language. *Developmental Psychology, 29*, 448–457.

Gloning, I., & Gloning, K. (1983). Aphasia in polyglots contribution to the dynamics of language disintegration as well as to the question of the localization of these impairments. In M. Paradis (Ed.), *Readings on aphasia in bilinguals and polyglots* (pp. 681–716). Montreal: Marcel Didier.

Gobet, F. (2006). Adriaan de Groot: Marriage of two passions. *International Computer Game Association Journal, 29*, 236–244.

Gobet, F., Lane, P. C. R., Croker, S., Cheng, P., Jones, G., Oliver, I., & Pine, J. M. (2001). Chunking mechanisms in human learning. *Trends in Cognitive Sciences, 5*, 236–243.

Gogtay, N., Giedd, J. N., Lusk, L., Hayashi, K. M., Greenstein, D., Vaituzis, A. C., ... Thompson, P. M. (2004). Dynamic mapping of human cortical development during childhood through early adulthood. *Proceedings of the National Academy of Sciences of the United States of America, 101*, 8174–8179.

Gollan, T. H., Montoya, R. I., Cera, C., & Sandoval, T. C. (2008). More use almost always means a smaller frequency effect: Aging, bilingualism, and the weaker links hypothesis. *Journal of Memory and Language, 58*, 787–814.

Gollan, T. H., Montoya, R. I., Fennema-Notestine, C., & Morris, S. K. (2005). Bilingualism affects picture naming but not picture classification. *Memory and Cognition, 33*, 1220–1234.

Gollan, T. H., Montoya, R. I., & Werner, G. A. (2002). Semantic and letter fluency in Spanish-English bilinguals. *Neuropsychology, 16*, 562–576.

Gomez-Tortosa, E., Martin, E., Gaviria, M., Charbel, F., & Ausman, J. (1995). Selective deficit of one language in a bilingual patient following surgery in the left perisylvian area. *Brain and Language, 48*, 320–325.

Gou, Z., Choudhury, N., & Benasich, A. A. (2010). Resting frontal gamma power at 16, 24 and 36 months predicts individual differences in language and cognition at 4 and 5 years. *Behavioural Brain Research, 220*, 263–270.

Grainger, J. (1990). Word frequency and neighborhood frequency effects in lexical decision and naming. *Journal of Memory and Language, 29*, 228–244.

Grainger, J., & Beauvillain, C. (1988). Associative priming in bilinguals: Some limits of interlingual facilitation effects. *Canadian Journal of Psychology, 42*, 261–273.

Graves, R. E. (1997). The legacy of the Wernicke-Lichtheim model. *Journal of the History of the Neurosciences, 6*, 3–20.

Gray, R. (2004). Attending to the execution of a complex sensorimotor skill: expertise differences, choking, and slumps. *Journal of Experimental Psychology Applied, 10*, 42–54.

Gray, S., Watts, S., Debicki, D., & Hore, J. (2006). Comparison of kinematics in skilled and unskilled arms of the same recreational baseball players. *Journal of Sports Sciences, 24*, 1183–1194.

Green, D. W., & Abutalebi, J. (2008). Understanding the link between bilingual aphasia and language control. *Journal of Neurolinguistics, 21*, 558–576.

Guion, S. G. (2003). The vowel systems of Quichua-Spanish bilinguals: An investigation into age of acquisition effects on the mutual influence of the first and second languages. *Phonetica, 60*, 98–128.

Harley, B., & Wang, W. (1997). The critical period hypothesis: Where are we now? In A. M. B. De Groot & J. F. Kroll (Eds.), *Tutorials in bilingualism: Psycholinguistic perspectives* (pp. 19–52). Hillsdale, NJ: Erlbaum.

Hart, B., & Risley, T. R. (1995). *Meaningful differences in the everyday experiences of young American children*. Baltimore, MD: Brookes.

Harwerth, R., Smith, E., Duncan, G., Crawford, M., & von Noorden, G. (1986). Multiple sensitive periods in the development of the primate visual system. *Science, 232*, 235–238.

Heath, S. B. (1983). *Ways with words: language, life, and work in communities and classrooms.* Cambridge, UK; New York: Cambridge University Press.

Heath, S. B. (1990). The children of Trackton's children: Spoken and written language in social change. In J. W. Stigler, R. A. Shweder, & G. Herdt (Eds.), *Cultural psychology: Essays on comparative human development* (pp. 496–519). Cambridge: Cambridge University Press.

Heim, S., Opitz, B., & Friederici, A. D. (2002). Broca's area in the human brain is involved in the selection of grammatical gender for language production: Evidence from event-related functional magnetic resonance imaging. *Neuroscience Letters, 328*, 101–104.

Heredia, R. R. (1997). Bilingual memory and hierarchical models: A case for language dominance. *Current Directions in Psychological Science, 6*, 34–39.

Hernandez, A., Li, P., & MacWhinney, B. (2005). The emergence of competing modules in bilingualism. *Trends in Cognitive Sciences, 9*, 220–225.

Hernandez, A. E., Bates, E., & Avila, L. X. (1994). Sentence interpretation in Spanish-English bilinguals: What does it mean to be in-between. *Applied Psycholinguistics, 15*, 417–466.

Hernandez, A. E., Bates, E., & Avila, L. X. (1996). Processing across the language boundary: A cross modal priming study of Spanish-English bilinguals. *Journal of Experimental Psychology: Learning, Memory, and Cognition, 22*, 846–864.

Hernandez, A. E., Dapretto, M., Mazziotta, J., & Bookheimer, S. (2001). Language switching and language representation in Spanish-English bilinguals: An fMRI study. *NeuroImage, 14*, 510–520.

Hernandez, A. E., & Kohnert, K. (1999). Aging and language switching in bilinguals. *Aging, Neuropsychology and Cognition, 6*, 69–83.

Hernandez, A. E., Kotz, S. A., Hofmann, J., Valentin, V. V., Dapretto, M., & Bookheimer, S. Y. (2004). The neural correlates of grammatical gender decisions in Spanish. *Neuroreport, 15*, 863–866.

Hernandez, A. E., Martinez, A., & Kohnert, K. (2000). In search of the language switch: An fMRI study of picture naming in Spanish-English bilinguals. *Brain and Language, 73*, 421–431.

Hernandez, A. E., & Reyes, I. (2002). Within- and between-language priming differ: evidence from repetition of pictures in Spanish-English bilinguals. *Journal of Experimental Psychology: Learning, Memory, and Cognition, 28*, 726–734.

Hernandez-Chavez, E., Burt, M., & Dulay, H. (1978). Language dominance and proficiency testing: Some general considerations. *NABE Journal, 3*, 41–54.

Herschmann, H., & Pötzl, O. (1983). Observations on aphasia in polyglots. In M. Paradis (Ed.), *Readings on aphasia in bilinguals and polyglots* (pp. 148–154). Montreal: Marcel Didier.

Hervais-Adelman, A. G., Moser-Mercer, B., & Golestani, N. (2011). Executive control of language in the bilingual brain: integrating the evidence from neuroimaging to neuropsychology. *Frontiers in Psychology, 2*, 234.

Hess, E. (2008). *Nim Chimpsky: The chimp who would be human*. New York: Bantam Books.

Hickok, G., & Poeppel, D. (2004). Dorsal and ventral streams: A framework for understanding aspects of the functional anatomy of language. *Cognition, 92*, 67–99.

Hickok, G., & Poeppel, D. (2007). The cortical organization of speech processing. *Nature Reviews: Neuroscience, 8*, 393–402.

Hilchey, M., & Klein, R. (2011). Are there bilingual advantages on nonlinguistic interference tasks? Implications for the plasticity of executive control processes. *Psychonomic Bulletin and Review, 18*, 625–658.

Hoff, E., Laursen, B., & Tardif, T. (2002). Socioeconomic status and parenting. In M. H. Bornstein (Ed.), *Handbook of parenting: Ecology and biology of parenting.* (Vol. II, pp. 161–188). Mahwah, NJ: Erlbaum.

Holtzheimer, P., Fawaz, W., Wilson, C., & Avery, D. (2005). Repetitive transcranial magnetic stimulation may induce language switching in bilingual patients. *Brain Lang, 94*, 274–277.

Hsieh, S., & Allport, A. (1994). Shifting attention in a rapid visual search paradigm. *Perceptual and Motor Skills, 79*, 315–335.

Huang, Z., Kirkwood, A., Pizzorusso, T., Porciatti, V., Morales, B., Bear, M.,... Tonegawa, S. (1999). BDNF regulates the maturation of inhibition and the critical period of plasticity in mouse visual cortex. *Cell, 98*.

Hubel, D. H., & Wiesel, T. N. (1963). Receptive fields of cells in striate cortex of very young, visually inexperienced kittens. *Journal of Neurophysiology, 26*, 994–1002.

Hughlings Jackson, J. (1876). Case of large cerebral tumour without optic neuritis and left hemiplegia and imperception. *Royal London Ophthalmic Hospital Reports, 8*, 434.

Huttenlocher, P. R. (1990). Morphometric study of human cerebral cortex development. *Neuropsychologia, 28*, 517–527.

Huttenlocher, P. R. (1994). Synaptogenesis in the human cerebral cortex. In G. Dawson & K. W. Fischer (Eds.), *Human Behavior and the Developing Brain* (pp. 137–152). New York: Guilford Press.

Huttenlocher, P. R., & Dabholkar, A. S. (1997). Regional differences in synaptogenesis in human cerebral cortex. *Journal of Comparative Neurology, 387*, 167–178.

Huttenlocher, P. R., & de Courten, C. (1987). The development of synapses in striate cortex of man. *Human Neurobiology, 6*, 1–9.

Huttenlocher, P. R., De Courten, C., Garey, L. J., & Van der Loos, H. (1982a). Synaptic development in human cerebral cortex. *International Journal of Neurology, 16–17*, 144–154.

Huttenlocher, P. R., de Courten, C., Garey, L. J., & Van der Loos, H. (1982b). Synaptogenesis in human visual cortex—evidence for synapse elimination during normal development. *Neuroscience Letters, 33*, 247–252.

Hyltenstam, K., Abrahamsson, N., Doughty, C. J., & Long, M. H. (2003). Maturational constraints in SLA. In C. J. Doughty & M. H. Long (Eds.), *The handbook of second language acquisition* (pp. 539–588). Malden, MA: Blackwell.

Hyltenstam, K., Bylund, E., Abrahamsson, N., & Park, H.-S. (2009). Dominant-language replacement: The ease of international adoptees. *Bilingualism: Language and Cognition, 12*, 121–140.

Issa, N., Trachtenberg, J., Chapman, B., Zahs, K., & Stryker, M. (1999). The critical period for ocular dominance plasticity in the ferret's visual cortex. *Journal of Neuroscience, 19*, 6965–6978.

Ivanova, I., & Costa, A. (2008). Does bilingualism hamper lexical access in speech production? *Acta Psychologica, 127*, 277–288.

Jackson, G. M., Swainson, R., Cunnington, R., & Jackson, S. R. (2001). ERP correlates of executive control during repeated language switching. *Bilingualism: Language and Cognition, 4*, 169–178.

Jersild, A. T. (1927). Mental set and shift. *Archives of Psychology, 89*, 5–82.

Jimura, K., & Braver, T. S. (2011). Age-related shifts in brain activity dynamics during task switching. *Cerebral Cortex, 20*, 1420–1431.

Johnson, J. S., & Newport, E. L. (1989). Critical period effects in second language learning: The influence of maturational state on the acquisition of English as a second language. *Cognitive Psychology, 21*, 60–99.

Johnson, M. H. (2005). Subcortical face processing. *Nature Reviews: Neuroscience, 6*, 766–774.

Johnson, M. H., & Munakata, Y. (2005). Processes of change in brain and cognitive development. *Trends in Cognitive Sciences, 9*, 152–188.

Jonides, J., Smith, E. E., Koeppe, R. A., & Awh, E. (1993). Spatial working memory in humans as revealed by PET. *Nature, 363*, 623–625.

Just, M. A., Carpenter, P. A., Keller, T. A., Eddy, W. F., & Thulborn, K. R. (1996). Brain activation modulated by sentence comprehension. *Science, 274*, 114–116.

Justice, L. M., Mashburn, A., Pence, K. L., & Wiggins, A. (2008). Experimental evaluation of a preschool language curriculum: Influence on children's expressive language skills. *Journal of Speech, Language, and Hearing Research, 51*, 983–1001.

Kaas, J. H. (2004). Somatosensory system. In G. Paxinos & J. K. Mai (Eds.), *The Human Nervous System* (2nd ed., pp. 1059–1092). New York: Elsevier Academic Press.

Kanwisher, N., McDermott, J., & Chun, M. M. (1997). The fusiform face area: A module in human extrastriate cortex specialized for face perception. *Journal of Neuroscience, 17*, 4302–4311.

Kauders, O. (1983). On polyglot responses in a sensory aphasia. In M. Paradis (Ed., Trans.), *Readings on aphasia in bilinguals and polyglots* (pp. 286–300). Montreal: Marcel Didier. Originally published in 1929.

Kaushanskaya, M., & Marian, V. (2009). Bilingualism reduces native-language interference during novel-word learning. *Journal of Experimental Psychology: Learning, Memory, and Cognition, 35*, 829–835.

Keele, S. W. (1968). Movement control in skilled motor performance. *Psychological bulletin, 70*(6), 387–403.

Kim, K. H. S., Relkin, N. R., Lee, K.-M., & Hirsch, J. (1997). Distinct cortical areas associated with native and second languages. *Nature, 388*, 171–174.

Kimble, G. A., & Perlmutter, L. C. The problem of volition. *Psychological Review*, 1970, *77*, 361–384.

King, J.W., & Kutas, M. 1995. Who did what and when? Using word- and clause-level ERPs to monitor working memory usage in reading. *Journal of Cognitive Neuroscience, 7*(3), 376–395.

Kohnert, K., & Bates, E. (2002). Balancing Bilinguals II: Lexical Comprehension and Cognitive Processing in Children Learning Spanish and English. *Journal of Speech, Language, and Hearing Research, 45*, 347–359.

Kohnert, K., Bates, E., & Hernandez, A. E. (1999). Balancing bilinguals: Lexical-semantic production and cognitive processing in children learning Spanish and English. *Journal of Speech, Language and Hearing Research, 42*, 1400–1413.

Kolb, B., & Whishaw, I. Q. (1985). *Fundamentals of human neuropsychology*. New York: Freeman.

Kolers, P. A. (1966). Interlingual facilitation of short-term memory. *Journal of Verbal Learning and Verbal Behavior, 5*, 314–319.

Konishi, T. (1993). The semantics of grammatical gender: A cross-cultural study. *Journal of Psycholinguistic Research, 22*, 519–534.

Kotz, S. A., & Elston-Guettler, K. (2004). The role of proficiency on processing categorical and associative information in the L2 as revealed by reaction times and event related brain potentials. *Journal of Neurolinguistics, 17*, 215–235.

Kousaie, S., & Phillips, N. A. (2012). Conflict monitoring and resolution: are two languages better than one? Evidence from reaction time and event-related brain potentials. *Brain Research, 1446*, 71–90.

Kroll, J. F., & de Groot, A. M. B. (1997). Lexical and conceptual memory in the bilingual: Mapping form to meaning in two languages. In A. M. B. de Groot & J. F. Kroll (Eds.), *Tutorials in bilingualism: Psycholinguistic perspectives.* (pp. 169–199). Mahwah, NJ: Erlbaum.

Kroll, J. F., & de Groot, A. M. B. (2005). *Handbook of bilingualism: Psycholinguistic approaches.* New York: Oxford University Press.

Kroll, J. F., & Stewart, E. (1994). Category inference in translation and picture naming: Evidence for asymmetric connections between bilingual memory representations. *Journal of Memory and Language, 33*, 149–174.

Kroll, J. F., & Tokowicz, N. (2005). Models of bilingual representation and processing: Looking back and to the future. In J. F. Kroll & A. M. B. DeGroot (Eds.), *Handbook of bilingualism: Psycholinguistic approaches.* (pp. 531–533). New York: Oxford University Press.

Kroll, J. F., Tokowicz, N., & Nicol, J. L. (2001). The development of conceptual representation for words in a second language. In J. Nicol (Ed.), *One mind, two languages: Bilingual language processing.* (pp. 49–71). Oxford: Blackwell Publishing.

Kroll, J. F., Van Hell, J. G., Tokowicz, N., & Green, D. W. (2010). The Revised Hierarchical Model: A critical review and assessment. *Bilingualism: Language and Cognition, 13*, 373–381.

Kuffler, S. W. (1953). Discharge patterns and functional organization of mammalian retina. *Journal of Neurophysiology, 16*, 37–68.

Kuhl, P. K. (2004). Early language acquisition: Cracking the speech code. *Nature Reviews Neuroscience, 5*, 831–841.

Kuhl, P. K., Tsao, F. M., & Liu, H. M. (2003). Foreign-language experience in infancy: effects of short-term exposure and social interaction on phonetic learning. *Proceedings of the National Academy of Sciences of the United States of America, 100*, 9096–9101.

Kushalnagar, P., Hannay, H. J., & Hernandez, A. E. (2010). Bilingualism and attention: A study of balanced and unbalanced bilingual deaf users of American Sign Language and English. *Journal of Deaf Studies and Deaf Education, 15*, 263–273.

Kutas, M., Van Petten, C., & Besson, M. (1988). Event-related potential asymmetries during the reading of sentences. *Electroencephalography and Clinical Neurophysiology, 69*.

Lange, J., Pötzl, O., Brown, J. W., & Liepmann, H. (1988). *Agnosia and apraxia: Selected papers of Liepmann, Lange, and Pötzl.* Hillsdale, NJ: Erlbaum.

Lau, E. F., Phillips, C., & Poeppel, D. (2008). A cortical network for semantics: (de)constructing the N400. *Nature Reviews Neuroscience, 9*, 920–933.

Lebrun, Y. (1995). The study of bilingual aphasia: Pitres legacy. In M. Paradis (Ed.), *Aspects of bilingual aphasia* (pp. 11–21). London: Pergamon.

Lenneberg, A. (1967). *Biological foundations of language.* New York: John Wiley.

Li, P., Farkas, I., & MacWhinney, B. (2004). Early lexical development in a self-organizing neural networks. *Neural Networks, 17*, 1345–1362.

Linck, J. A., Kroll, J. F., & Sunderman, G. (2009). Losing access to the native language while immersed in a second language: Evidence for the role of inhibition in second-language learning. *Psychological Science, 20*, 1507–1515.

Linebarger, M. C., Schwartz, M. F., & Saffran, E. M. (1983). Sensitivity to grammatical structure in so-called agrammatic aphasics. *Cognition, 13*, 361–392.

Lorch, M. P. (2009). History of aphasia: Multiple languages, memory, and regression: An examination of Ribot's Law. *Aphasiology, 23*, 643–654.

MacNamara, J., Krauthammer, M., & Bolgar, M. (1968). Language switching in bilinguals as a function of stimulus and response uncertainty. *Journal of Experimental Psychology, 78*, 208–215.

MacNamara, J., & Kushnir, S. (1971). Linguistic independence of bilinguals: The input switch. *Journal of Verbal Learning and Verbal Behavior, 10*, 480–487.

MacWhinney, B. (2004). A unified model of language acquisition. In J. F. Kroll and A. M. B. de Groot (Eds.), *Handbook of bilingualism: Psycholinguistic approaches* (pp. 49–67). New York: Oxford University Press.

MacWhinney, B., & Bates, E. (1978). Sentential devices for conveying giveness and newness: A cross-cultural development study. *Journal Verbal Learning and Verbal Behavior, 17*, 539–558.

MacWhinney, B., Plelh, C., & Bates, E. (1985). The development of sentence interpretation in Hungarian. *Cognitive Psychology, 17*, 178–209.

Madden, D. J., Costello, M. C., Dennis, N. A., Davis, S. W., Shepler, A. M., Spaniol, J., . . . Cabeza, R. (2010). Adult age differences in functional connectivity during executive control. *NeuroImage, 52*, 643–657.

Maurer, D., Mondloch, C. J., Robertson, L. C., & Sagiv, N. (2005). Neonatal Synesthesia: A reevaluation. In L. C. Robertson and N. Sagiv (Eds.), *Synesthesia: Perspectives from cognitive neuroscience* (pp. 193–213). New York: Oxford University Press.

Mayberry, R. I. (2007). When timing is everything: Age of first-language acquisition effects on second-language learning. *Applied Psycholinguistics, 28*, 537–549.

Mayberry, R. I., & Lock, E. (2003). Age constraints on first versus second language acquisition: Evidence for linguistic plasticity and epigenesis. *Brain and Language, 87*, 369–383.

Mayberry, R. I., Lock, E., & Kazmi, H. (2002). Linguistic ability and early language exposure. *Nature, 417*, 38.

McClelland, J. L., & Rumelhart, D. E. (1988). *Explorations in parallel distributed processing: A handbook of models, programs, and exercises.* Cambridge, MA: MIT Press.

McGivern, R. F., Andersen, J., Byrd, D., Mutter, K. L., & Reilly, J. (2002). Cognitive efficiency on a match to sample task decreases at the onset of puberty in children. *Brain and Cognition, 50*, 73–89.

McLaughlin, J., Osterhout, L., & Kim, A. (2004). Neural correlates of second-language word learning: Minimal instruction produces rapid change. *Nature Neuroscience, 7*, 703–704.

Mechelli, A., Crinion, J. T., Noppeney, U., O'Doherty, J., Ashburner, J., Frackowiak, R. S., & Price, C. J. (2004). Neurolinguistics: Structural plasticity in the bilingual brain. *Nature, 431*, 757.

Mehler, J., Jusczyk, P., Lambertz, G., & Halsted, N. (1988). A precursor of language acquisition in young infants. *Cognition, 29*, 143–178.

Meisel, J. M. (2001). The simultaneous acquisition of two first languages: Early differentiation and subsequent development of grammars. In J. Cenoz & F. Genesee (Eds.), *Trends in bilingual acquisition* (pp. 11–42). Amsterdam: Benjamins.

Meschyan, G., & Hernandez, A. (2002). Age of acquisition and word frequency: determinants of object-naming speed and accuracy. *Memory and Cognition, 30*, 262–269.

Meschyan, G., & Hernandez, A. E. (2006). Impact of language proficiency and orthographic transparency on bilingual word reading: An fMRI investigation. *NeuroImage, 29*, 1135–1140.

Meuter, R. F. I., & Allport, A. (1999). Bilingual language switching in naming: Asymmetrical costs of language selection. *Journal of Memory and Language, 40*.

Minkowski, M. (1983). A clinical contribution to the study of polyglot aphasia especially with respect to Swiss-German. In M. Paradis (Ed.), *Readings on aphasia in bilinguals and polyglots* (pp. 205–232). Montreal: Marcel Didier.

Monaghan, J., & Ellis, A. W. (2002). Age of acquisition and the completeness of phonological representations. *Reading and Writing, 15*, 759–788.

Mondloch, C. J., Le Grand, R., Maurer, D., Gauthier, I., Tarr, M. J., & Bub, D. (2010). Development of expertise in face recognition. In I. Gauthier, M. J. Tarr, & D. Bub (Eds.), *Perceptual expertise: Bridging brain and behavior* (pp. 67–106). New York: Oxford University Press.

Moreno, E. M., Federmeier, K. D., & Kutas, M. (2002). Switching languages, switching Palabras (words): An electrophysiological study of code switching. *Brain and Language, 80*, 188–207.

Morgan, J. L., & Demuth, K. (1996a). Signal to syntax: An overview. In J. L. Morgan & K. Demuth (Eds.), *Signal to syntax: Bootstrapping from speech to grammar in early acquisition* (pp. 1–22). Mahwah, NJ: Erlbaum.

Morgan, J. L., & Demuth, K. (Eds.). (1996b). *Signal to syntax: Bootstrapping from speech to grammar in early acquisition*. Mahwah, NJ: Erlbaum.

Morrison, C. M., Chappell, T. D., & Ellis, A. W. (1997). Age of acquisition norms for a large set of object names and their relation to adult estimates and other variables. *Quarterly Journal of Experimental Psychology: Human Experimental Psychology, 50A*, 528–559.

Morrison, C. M., & Ellis, A. W. (1995). Roles of word frequency and age of acquisition in word naming and lexical decision. *Journal of Experimental Psychology: Learning, Memory, and Cognition, 21*, 116–133.

Morton, J., & Johnson, M. H. (1991). Conspec and conlern: A two process theory of infant face recognition. *Psychological Review, 98*, 164–181.

Morton, J. B., & Harper, S. N. (2007). What did Simon say? Revisiting the bilingual advantage. *Developmental Science, 10*, 719–726.

Neubauer, A. C., & Fink, A. (2009). Intelligence and neural efficiency. *Neuroscience and Biobehavioral Reviews, 33*, 1004–1023.

Norcia, A. M. (1996). Abnormal motion processing and binocularity: Infantile esotropia as a model system for effects of early interruptions of binocularity. *Eye, 10*, 259–265.

O'Hare, E. D., Lu, L. H., Houston, S. M., Bookheimer, S. Y., & Sowell, E. R. (2008). Neurodevelopmental changes in verbal working memory load-dependency: an fMRI investigation. *NeuroImage, 42*, 1678–1685.

Oh, J. S., Au, T. K.-F., & Jun, S.-A. (2010). Early childhood language memory in the speech perception of international adoptees. *Journal of Child Language, 37*, 1123–1132.

Oh, J. S., Jun, S.-A., Knightly, L. M., & Au, T. K.-F. (2003). Holding on to childhood language memory. *Cognition, 86*, B53–b64.

Ojemann, G. A., & Whitaker, H. A. (1978). The bilingual brain. *Archives of Neurology, 35*, 409–412.

Oller, J. W. (1978). The language factor in the evaluation of bilingual education. In J. E. Alatis (Ed.), *Georgetown University Round Table on Languages and Linguistics*. Washington, DC: Georgetown University Press.

Oller, J. W. (1979). *Language tests at school: A pragmatic approach*. New York: Longman.

Olson, C., & Freeman, R. (1980). Profile of the sensitive period for monocular deprivation in kittens. *Experimental Brain Research, 39*, 17–21.

Ostergaard, A. L. (1998). The effects on priming of word frequency, number of repetitions, and delay depend on the magnitude of priming. *Memory and Cognition, 26*, 40–60.

Osterhout, L., & Holcomb, P. J. (1992). Event-related brain potentials elicited by syntactic anomaly. *Journal of Memory and Language, 31*, 785–806.

Osterhout, L., Holcomb, P. J., & Swinney, D. A. (1994). Brain potentials elicited by garden-path sentences: Evidence of the application of verb information during parsing. *Journal of Experimental Psychology: Learning, Memory, and Cognition, 20*, 786–803.

Pakulak, E., & Neville, H. J. (2010). Proficiency differences in syntactic processing of monolingual native speakers indexed by event-related potentials. *Journal of Cognitive Neuroscience, 22*, 2728–2744.

Pakulak, E., & Neville, H.J. (2011). Maturational constraints on the recruitment of early processes for syntactic processing. *Journal of Cognitive Neuroscience, 23*, 2752–2765. Epub ahead of print October 22, 2010.

Pallier, C., Dehaene, S., Poline, J. B., LeBihan, D., Argenti, A. M., Dupoux, E., & Mehler, J. (2003). Brain imaging of language plasticity in adopted adults: can a second language replace the first? *Cerebral Cortex, 13*, 155–161.

Papagno, C., & Vallar, G. (1995). Verbal short-term memory and vocabulary learning in polyglots. *Quarterly Journal of Experimental Psychology A: Human Experimental Psychology, 48A*, 98–107.

Paradis, J., & Genesee, F. (1996). Syntactic acquisition in bilingual children: Autonomous or interdependent? *Studies in Second Language Acquisition, 18*, 1–25.

Paradis, M. (2009). *Declarative and procedural determinants of second languages*. Amsterdam; Philadelphia: Benjamins.

Penfield, W. (1950). The cerebral cortex and the mind of man. In P. Laslett (Ed.), *The physical basis of mind* (pp. 56–64). Oxford: Basil Blackwell.

Penfield, W. (1965). Conditioning the uncommitted cortex for language learning. *Brain, 88*, 787–798.

Penfield, W., & Boldrey, E. (1937). Somatic motor and sensory representation in the cerebral cortex of man as studied by electrical stimulation. *Brain, 60*, 389–443. doi: 10.1093/brain/60.4.389

Penfield, W., & Roberts, L. (1959). *Speech and brain mechanisms*. Princeton, NJ: Princeton University Press.

Perani, D., Dehaene, S., Grassi, F., Cohen, L., Cappa, S., Dupoux, E., ... Mehler, J. (1996). Brain processing of native and foreign languages. *Neuroreport, 7*, 2439–2444.

Perani, D., Paulesu, E., Galles, N. S., Dupoux, E., Dehaene, S., Bettinardi, V., ... Mehler, J. (1998). The bilingual brain: Proficiency and age of acquisition of the second language. *Brain, 121*, 1841–1852.

Petitto, L. A., Berens, M. S., Kovelman, I., Dubins, M. H., Jasinska, K., & Shalinsky, M. (2012). The "Perceptual Wedge Hypothesis" as the basis for bilingual babies' phonetic processing advantage: New insights from fNIRS brain imaging. *Brain and Language, 121*, 130–143.

Piaget, J. (1954). *The construction of reality in the child* (M. Cook, Trans.). New York: Basic Books.

Piaget, J. (1970). *Genetic epistemology* (E. Duckworth, Trans.). New York: Columbia University Press.

Pinker, S. (1994). *The language Instinct.* New York: William Morrow.

Piske, T., MacKay, I. R. A., & Flege, J. E. (2001). Factors affecting degree of foreign accent in an L2: A review. *Journal of Phonetics, 29*, 191–215.

Poldrack, R. A., Wagner, A. D., Prull, M. W., Desmond, J. E., Glover, G. H., & Gabrieli, J. D. (1999). Functional specialization for semantic and phonological processing in the left inferior prefrontal cortex. *NeuroImage, 10*, 15–35.

Potter, M. C., So, K.-f., von Eckardt, B., & Feldman, L. B. (1984). Lexical and conceptual representation in beginning and proficient bilinguals. *Journal of Verbal Learning and Verbal Behavior, 23*, 23–38.

Pötzl, O. (1983). Aphasia and mutilingualism. In M. Paradis (Ed.), *Readings on aphasia in bilinguals and polyglots* (pp. 301–316). Montreal: Marcel Didier.

Prat, C. S. (2011). The brain basis of individual differences in language comprehension abilities. *Language and Linguistics Compass, 5*, 635–649.

Prat, C. S., & Just, M. A. (2011). Exploring the neural dynamics underpinning individual differences in sentence comprehension. *Cerebral Cortex, 21*, 1747–1760. doi: 10.1093/cercor/bhq241

Price, C. J., & Devlin, J. T. (2003). The myth of the visual word form area. *NeuroImage, 19*, 473–481.

Price, C. J., & Devlin, J. T. (2011). The interactive account of ventral occipitotemporal contributions to reading. *Trends in Cognitive Sciences, 15*, 246–253.

Price, C. J., Green, D. W., & Von Studnitz, R. (1999). A functional imaging study of translation and language switching. *Brain, 122*, 2221–2235.

Proctor, R. W., & Dutta, A. (1995). *Skill acquisition and human performance.* Thousand Oaks, CA: Sage.

Provost, J.-S., Petrides, M., & Monchi, O. (2010). Dissociating the role of the caudate nucleus and dorsolateral prefrontal cortex in the monitoring of events within human working memory. *European Journal of Neuroscience, 32*, 873–880.

Reicher, G. M. (1969). Perceptual recognition as a function of meaningfulness of stimulus material. *Journal of Experimental Psychology, 81*, 275–280.

Reisberg, D. (2010). *Cognition: Exploring the science of the mind.* New York: Norton.

Reyes, I., & Hernandez, A. E. (2006). Sentence interpretation strategies in emergent bilingual children and adults. *Bilingualism: Language and Cognition, 9*, 51–69.

Ribot, T. (1881). *Les maladies de la mémoire* (2nd ed.). Paris: Baillere.

Ribot, T. (1882). *Diseases of memory: An essay in the positive psychology.* New York: Appleton.

Rodriguez-Fornells, A., Kramer, U. M., Lorenzo-Seva, U., Festman, J., & Munte, T. F. (2011). Self-assessment of individual differences in language switching. *Frontiers in Psychology, 2*, 388.

Rogers, R. D., & Monsell, S. (1995). Costs of a predictable switch between simple cognitive tasks. *Journal of Experimental Psychology: General, 124*, 207–231.

Rossi, S., Gugler, M. F., Friederici, A. D., & Hahne, A. (2006). The impact of proficiency on syntactic second-language processing of German and Italian: Evidence from event-related potentials. *Journal of Cognitive Neuroscience, 18*, 2030–2048.

Rowe, J. B., & Passingham, R. E. (2001). Working memory for location and time: Activity in prefrontal area 46 relates to selection rather than maintenance in memory. *NeuroImage, 14*, 77–86.

Rueda, M. R., Fan, J., McCandliss, B. D., Halparin, J. D., Gruber, D. B., Lercari, L. P., & Posner, M. I. (2004). Development of attentional networks in childhood. *Neuropsychologia, 42*, 1029–1040.

Rumelhart, D., & McClelland, J. L. (1986). *Parallel distributed processing: Explorations in the microstructure of cognition. Vol. 1: Foundations.* Cambridge, MA: MIT Press.

Saffran, J. R. (2001). Words in a sea of sounds: The output of infant statistical learning. *Cognition, 81*, 149–169.

Saffran, J. R., Aslin, R. N., & Newport, E. L. (1996). Statistical learning by 8-month-old infants. *Science, 274*, 1926–1928.

Sanz, C. (2000). Bilingual education enhances third language acquisition: Evidence from Catalonia. *Applied Psycholinguistics, 21*, 23–44.

Schafer, E. W. (1982). Neural adaptability: A biological determinant of behavioral intelligence. *International Journal of Neuroscience, 17*, 183–191.

Schlaggar, B. L., Brown, T. T., Lugar, H. M., Visscher, K. M., Miezin, F. M., & Petersen, S. E. (2002). Functional neuroanatomical differences between adults and school-age children in the processing of single words. *Science, 296*, 1476–1479.

Schlaug, G., Jancke, L., Huang, Y., Staiger, J. F., & Steinmetz, H. (1995). Increased corpus callosum size in musicians. *Neuropsychologia, 33*, 1047–1055.

Schreuder, R., & Weltens, B. (Eds.). (1993). *The bilingual lexicon.* Amsterdam: Benjamins.

Sergent, J., Ohta, S., & Macdonald, B. (1992). Functional neuroanatomy of face and object processing. *Brain, 115*, 15–36. doi: 10.1093/brain/115.1.15

Shinskey, J. L., & Munakata, Y. (2003). Are infants in the dark about hidden objects? *Developmental Science, 6*, 273–282.

Sholl, A., Sankaranarayanan, A., & Kroll, J. F. (1995). Transfer between picture naming and translation: A test of asymmetries in bilingual memory. *Psychological Science, 6*, 45–49.

Slobin, D. I. (1985). Crosslinguistic evidence for the language-making capacity. In D. I. Slobin (Ed.), *The crosslinguistic study of language acquisition, Vol. 1: The data; Vol. 2: Theoretical issues.* (pp. 1157–1256). Hillsdale, NJ: Erlbaum.

Smith, L. B., & Thelen, E. (1993). *A dynamic systems approach to development: applications.* Cambridge, Mass.: MIT Press.

Smith, L. B., & Thelen, E. (2003). Development as a dynamic system. *Trends in Cognitive Sciences, 7*, 343–348.

Soares, C., & Grosjean, F. (1984). Bilinguals in a monolingual and a bilingual speech mode: The effect on lexical access. *Memory and Cognition, 12*, 380–386.

Sowell, E. R., Peterson, B. S., Thompson, P. M., Welcome, S. E., Henkenius, A. L., & Toga, A. W. (2003). Mapping cortical change across the human life span. *Nature Neuroscience, 6*, 309–315.

Spector, A., & Biederman, I. (1976). Mental set and mental shift revisited. *American Journal of Psychology, 89*, 669–679.

Squire, L. R., & Zola, S. M. (1996). Structure and function of declarative and nondeclarative memory systems. *Proceedings of the National Academy of Sciences of the United States of America, 93*, 13515–13522.

Steinhauer, K., White, E. J., & Drury, J. E. (2009). Temporal dynamics of late second language acquisition: evidence from event-related brain potentials. *Second Language Research, 25*, 13–41.

Stengel, E., & Zelmanowicz, J. (1933). On polyglot motor aphasia. In M. Paradis (Ed.), *Readings on aphasia in bilinguals and polyglots* (pp. 356–375). Montreal: Marcel Didier.

Stout, D. (2010). The evolution of cognitive control. *Topics in Cognitive Science, 2*, 614–630.

Strayer, D. L., & Drews, F. A. (2004). Profiles in driver distraction: Effects of cell phone conversations on younger and older drivers. *Human Factors, 46*, 640–649.

Strayer, D. L., & Johnston, W. A. (2001). Driven to distraction: Dual-task studies of simulated driving and conversing on a cellular telephone. *Psychological Science, 12*, 462–466.

Swinney, D. A. (1979). Lexical access during sentence comprehension: (Re)consideration of context effects. *Journal of Verbal Learning and Verbal Behavior, 18*, 645–659.

Taylor, I. A., & Getzels, J. W. (1975). *Perspectives in creativity*. Oxford, UK: Aldine.

Tees, R. C., & Werker, J. F. (1984). Perceptual flexibility: Maintenance or recovery of the ability to discriminate non-native speech sounds. *Canadian Journal of Psychology/Revue Canadienne de Psychologie, 38*, 579–590.

Telkemeyer, S., Rossi, S., Koch, S. P., Nierhaus, T., Steinbrink, J., Poeppel, D., . . . Wartenburger, I. (2009). Sensitivity of newborn auditory cortex to the temporal structure of sounds. *Journal of Neuroscience, 29*, 14726–14733.

Terrace, H. S. (1987). *Nim*. New York: Columbia University Press.

Thompson, P. (1980). Margaret Thatcher: A new illusion. *Perception, 9*, 483–484.

Toga, A. W., Thompson, P. M., & Sowell, E. R. (2006). Mapping brain maturation. *Trends in Neurosciences, 29*, 148–159.

Trainor, L. J. (2005). Are there critical periods for musical development? *Developmental Psychobiology, 46*, 262–278.

Uddin, L. Q., Iacoboni, M., Lange, C., & Keenan, J. P. (2007). The self and social cognition: the role of cortical midline structures and mirror neurons. *Trends in Cognitive Sciences, 11*, 153–157.

Ukrainetz, T. A., & Blomquist, C. (2002). The criterion validity of four vocabulary tests compared to a language sample. *Child Language Teaching and Therapy, 18*, 59–78.

Valenzuela, M. J., & Sachdev, P. (2006a). Brain reserve and cognitive decline: A non-parametric systematic review. *Psychological Medicine, 36*, 1065–1073.

Valenzuela, M. J., & Sachdev, P. (2006b). Brain reserve and dementia: a systematic review. *Psychological Medicine, 36*, 441–454.

van Gelder, T. (1998). The dynamical hypothesis in cognitive science. *Behavioral and Brain Sciences, 21*, 615–628.

Van Hell, J. G., & Mahn, A. C. (1997). Keyword mnemonics versus rote rehearsal: Learning concrete and abstract foreign words by experienced and inexperienced learners. *Language Learning, 47*, 507–546.

van Heuven, W. J. B., Dijkstra, T., & Grainger, J. (1998). Orthographic neighborhood effects in bilingual word recognition. *Journal of Memory and Language, 39*, 458–483.

van Schaik, C. P., Deaner, R. O., & Merrill, M. Y. (1999). The conditions for tool use in primates: Implications for the evolution of material culture. *Journal of Human Evolution, 36*, 719–741.

Ventureyra, V. A. G., Pallier, C., & Yoo, H.-Y. (2004). The loss of first language phonetic perception in adopted Koreans. *Journal of Neurolinguistics, 17*, 79–91.

Verhaeghen, P. (2003). Aging and vocabulary score: A meta-analysis. *Psychology and Aging, 18*, 332–339.

Verhaeghen, P., Cerella, J., Bopp, K. L., & Basak, C. (2005). Aging and varieties of cognitive control: A review of meta-analyses on resistance to interference, coordination, and task switching, and an experimental exploration of age-sensitivity in the newly identified process of focus switching. In R. W. Engle et al. (Eds.), *Cognitive limitations in aging and psychopathology* (pp. 160–189). New York: Cambridge University Press.

Vitali, P., Tettamanti, M., Abutalebi, J., Ansaldo, A. I., Perani, D., Cappa, S. F., & Joanette, Y. (2010). Generalization of the effects of phonological training for anomia using structural equation modelling: A multiple single-case study. *Neurocase, 16*, 93–105.

von Studnitz, R., & Green, D. (1997). Lexical decision and language switching. *International Journal of Bilingualism, 1*, 3–24.

Wagner, A. D., Maril, A., Bjork, R. A., & Schacter, D. L. (2001). Prefrontal contributions to executive control: fMRI evidence for functional distinctions within lateral Prefrontal cortex. *NeuroImage, 14*, 1337–1347.

Wagner, A. D., Pare-Blagoev, E. J., Clark, J., & Poldrack, R. A. (2001). Recovering meaning: Left prefrontal cortex guides controlled semantic retrieval. *Neuron, 31*, 329–338.

Wagner, V., Woods, E. A., Beilock, S., & Hernandez, A. E. (April 2010). *Brain activity during effortful auditory retrieval differentiates expert and non-expert athletes.* Paper presented at the Annual Meeting of the Cognitive Neuroscience Society, San Francisco.

Wallace, D. F. (2006). Federer as religious experience. *The New York Times,* August 20.

Wartenburger, I., Heekeren, H. R., Abutalebi, J., Cappa, S. F., Villringer, A., & Perani, D. (2003). Early setting of grammatical processing in the bilingual brain. *Neuron, 37*, 159–170.

Watanabe, D., Savion-Lemieux, T., & Penhune, V. B. (2007). The effect of early musical training on adult motor performance: Evidence for a sensitive period in motor learning. *Experimental Brain Research, 176*, 332–340.

Weber-Fox, C., & Neville, H. J. (1996). Maturational constraints on functional specializations for language processing: ERP and behavioral evidence in bilingual speakers. *Journal of Cognitive Neuroscience, 8*, 231–256.

Wells, C. S. (1979). Influences of the home on language development. *Bristol Working Papers on Language.*

Werker, J. F., & Tees, R. C. (2005). Speech perception as a window for understanding plasticity and commitment in language systems of the brain. *Developmental Psychobiology, 46*, 233–234.

Wheeler, D. D. (1970). Processes in word recognition. *Cognitive Psychology, 1*, 59–85.

Wiesel, T. N., & Hubel, D. H. (1963a). Effects of visual deprivation on morphology and physiology of cells in the cats lateral geniculate body. *Journal of Neurophysiology, 26*, 978–993.

Wiesel, T. N., & Hubel, D. H. (1963b). Single-cell responses in striate cortex of kittens deprived of vision in one eye. *Journal of Neurophysiology, 26*, 1003–1017.

Wimmer, H., & Perner, J. (1983). Beliefs about beliefs: Representation and constraining function of wrong beliefs in young children's understanding of deception. *Cognition, 13*, 103–128.

Wong Fillmore, L. (1991). Second-language learning in children: A model of language learning in social context. In E. Bialystok (Ed.), *Language processing in bilingual children* (pp. 49–69). Cambridge: Cambridge University Press.

Wong, Y. K., Twedt, E., Sheinberg, D., & Gauthier, I. (2010). Does Thompson's Thatcher Effect reflect a face-specific mechanism? *Perception, 39*, 1125–1141.

Xue, G., Dong, Q., Jin, Z., Zhang, L., & Wang, Y. (2004). An fMRI study with semantic access in low proficiency second language learners. *Neuroreport, 15*, 791–796.

Yoshida, H., Tran, D. N., Benitez, V., & Kuwabara, M. (2011). Inhibition and adjective learning in bilingual and monolingual children. *Frontiers in Developmental Psychology, 2*, 210.

Young, A., & van de Wal, C. (1996). Charcot's case of impaired imagery. In C. Code (Ed.), *Classic cases in neuropsychology*. New York: Psychology Press.

Zhao, X., & Li, P. (2010). Bilingual lexical interactions in an unsupervised neural network model. *International Journal of Bilingual Education and Bilingualism, 13*, 505–524.

Index